Praise for *Synergistic Selection*

"In this wide-ranging and intelligent book, Peter Corning presents a grand view of the evolution of complexity, eloquently arguing that it is based on the cumulative effects of cooperation – on synergies among systems. He provides an inclusive synthesis of evolutionary theory, re-interpreting classical Darwinian and Lamarckian views in terms of evolving synergistic interactions, and he offers a new way of thinking about the major transitions in evolution and human evolution. Brave, well-written, and based on more than 30 years of deep reflection, Corning's vision stretches into the future of our species and suggests new ways of anticipating and facing it."

– Eva Jablonka, Professor, The Cohn Institute for the History and Philosophy of Science and Ideas, Tel-Aviv University and co-author of (among others) *Evolution in Four Dimensions*

"The concept of society as an organism has a long history as a metaphor but only very recently has been placed on a solid scientific foundation. The implications for public policy are transformative and Peter Corning's *Synergistic Selection* provides an excellent guide for the general public and policy experts of all stripes."

– David Sloan Wilson, SUNY Distinguished Professor, Binghamton University, President, Evolution Institute, and author of (among others) *Does Altruism Exist?*

"Evolutionary biology is in the middle of its most exciting intellectual transformation since the publication of *The Origin of the Species*. Drawing on contemporary ideas from epigenetics, multi-level selection, bioeconomics, and more, Peter Corning has given us a book that not only expertly summarizes the current state of the art in evolutionary biology, it also points the way to a compelling new understanding of how we arrived at our place in the natural world."

– John M. Gowdy, Professor of Economics and Professor of Science and Technology Studies, Rensselaer Polytechnic Institute and co-author of *Paradise For Sale*

"In his new book *Synergistic Selection*, Peter Corning explains in an elegant and detailed way the rise of complexity in living systems over time and the major transitions in evolution. The book moves beyond neo-Darwinism and challenges the traditional selectionist approach, focusing on the role of additional evolutionary mechanisms associated with functional synergies and the emergence of evolutionary novelties."

– Francisco Carrapiço, Professor of Biology, University of Lisbon, Portugal

"My introduction to evolution was all about competition, population genetics, linear equations. Lamarck as idiot, Wynne-Edwards mocked. But on reading Waddington, Bohm and Heisenberg, doubts soon emerged. 50 years later we are in the midst of a major paradigm shift in which Peter Corning has been a pioneer. His book offers a wonderful overview of the new evolution landscape. Be gone intelligent designers and blind watchmakers – let the light of synergy in!"

– Dick Vane-Wright, Honorary Professor of Taxonomy, University of Kent, UK, Scientific Associate, the Natural History Museum, London, and editor of *The Role of Behaviour in Evolution*

"Synergy, selection, and emergence are central concepts in relation to complex systems. As we enter an era of ever smarter machines, and ever more powerful technologies, never have these phenomena been more important. Few have thought as deeply about them as Peter Corning. You could not find a better place to start an inquiry into the evolving science of the future of cooperation in a complex world."

– John Smart, Founder, Acceleration Studies Foundation and author of *The Foresight Guide*

"*Synergistic Selection* is an important contribution to our understanding of evolution… An essential read at the intersection of science, inspiration, and sustainability."

– Michael Dowd, author of *Thank God for Evolution* and host of "The Future Is Calling Us to Greatness"

SYNERGISTIC SELECTION

How Cooperation Has Shaped Evolution and the Rise of Humankind

SYNERGISTIC SELECTION

How Cooperation Has Shaped Evolution and the Rise of Humankind

Peter Corning

Institute for the Study of Complex Systems, USA

NEW JERSEY • LONDON • SINGAPORE • BEIJING • SHANGHAI • HONG KONG • TAIPEI • CHENNAI • TOKYO

Published by

World Scientific Publishing Co. Pte. Ltd.
5 Toh Tuck Link, Singapore 596224
USA office: 27 Warren Street, Suite 401-402, Hackensack, NJ 07601
UK office: 57 Shelton Street, Covent Garden, London WC2H 9HE

Library of Congress Cataloging-in-Publication Data
Names: Corning, Peter A., 1935– author.
Title: Synergistic selection : how cooperation has shaped evolution and the
rise of humankind / Peter Corning, Institute for the Study of Complex Systems, USA.
Description: New Jersey : World Scientific, 2018. | Includes bibliographical references and index.
Identifiers: LCCN 2017038819| ISBN 9789813230934 (hardcover : alk. paper) |
ISBN 9813230932 (hardcover : alk. paper) | ISBN 9789813234604 (pbk. : alk. paper) |
ISBN 9813234601 (pbk. : alk. paper)
Subjects: LCSH: Evolution (Biology) | Human evolution. | Social evolution. |
Synergetics. | Natural selection.
Classification: LCC QH366.2 .C733 2017 | DDC 576.8--dc23
LC record available at https://lccn.loc.gov/2017038819

British Library Cataloguing-in-Publication Data
A catalogue record for this book is available from the British Library.

For any available supplementary material, please visit
http://www.worldscientific.com/worldscibooks/10.1142/10732#t=suppl

Printed in Singapore

For Susan

Honoring a debt that can never be repaid

Foreword

In the Introduction to my original 1983 book, *The Synergism Hypothesis,* I drew the following conclusion about this new theory:

> Mindful of Albert Einstein's observation that "a theory is all the more impressive the greater is the simplicity of its premises, the more different are the kinds of things it relates and the more extended its range of applicability," I have come to believe that it is both possible and appropriate to reduce certain fundamental aspects of the evolutionary process, in nature and human societies alike, to a unifying theoretical framework...Equally important, my theory has had a life of its own…As the work proceeded, certain ideas emerged and certain connections were made. In the end, I have been impelled to follow where the theory led me. Only time will tell whether these connections were justified.

The subsequent history of the Synergism Hypothesis amply confirms Shakespeare's famous metaphor about how there are tides in human affairs, and that good timing can make all the difference.[i] In 1983, this theory faced a strong adverse tide. My book had the misfortune to appear during the heyday of the "selfish gene" model of evolution, at a time when there was a fixation with kin selection theory in sociobiology. (Much more

---

[i] The original version, from Shakespeare's *Julius Caesar*, reads as follows: "There is a tide in the affairs of men, which taken at the flood leads on to fortune; omitted, all the voyage of their life is bound in shallows and in miseries."

on this later on in the book.) The simple idea that functional synergy of various kinds might have played a major causal role in shaping the trajectory of evolution (in effect, an economic theory of biological complexity) seemed far-fetched and quixotic to many other scholars at the time. As one critic politely complained, "I do not understand it."

Thus, over the past 30-odd years I have viewed myself as being rather like a trial lawyer who must plead the same case repeatedly through various courts of law – to the degree that one old friend warned me against "running the subject into the ground." Fortunately, it seems that a more favorable tide is now running in evolutionary biology, and the judges are much more friendly. Indeed, one anonymous reviewer of a recent journal article related to this book commented waspishly that there doesn't seem to be anything new here: "It's rather obvious – a truism."

Over the years, I have accumulated many personal debts to various mentors, helpful colleagues, editors, and even some critics. Among these, I would especially include (roughly in chronological order) Jim Crown, Gerry McClearn, Theodosius Dobzhansky, Ernst Mayr, Cynthia Merman, Ken Boulding, John Maynard Smith, Eörs Szathmáry, Mike Ghiselin, Connie Barlow and Michael Dowd, Roger Masters, John Paul Scott, Tony Trewavas, Janet Landa, John Gowdy, Geoff Hodgson, Jack Hirschleifer, Steve Kline, Anatol Rapoport, Lynn Margulis, Ward Cooper, Iver Mysterud, Chris Boehm, David Sloan Wilson, Pete Richerson, Ulrich Witt, Christie Henry, Francisco Carrapiço, Dick Vane Wright, Herb Gintis, Stu Kauffman, Rollin Odell, Denis Noble, Bert Leigh, Kevin Laland, Klaus Jaffe, Pat Bateson, and Daniel McShea, along with countless anonymous reviewers over the years. My apologies to anyone I might have forgotten to mention. I am also immensely grateful for the consummate skills and many years of devoted service by my research associate, Marianne Siroker, and I owe a huge debt – of many kinds – to my wife Susan and to my family.

Contents

Chapter 1

Explaining Complexity

If nothing in biology makes sense except in the light of evolution – to paraphrase the pioneering twentieth century biologist Theodosius Dobzhansky – it's equally true that nothing about the evolution of biological complexity makes sense except in the light of synergy.[1]

Natural selection is a clever and immensely useful concept. It has provided us with a general theoretical framework for understanding how life has evolved over the past 3.8 billion years or so, and it is probably one of the best-known ideas in all of science (although it's often misunderstood – more on this later on).

However, Darwin's theory does not provide an explanation for the rise of biological complexity – one of the most consequential trends in the history of life on Earth. The trajectory of biological "complexification" – from primitive one-celled life forms to intricate eukaryotes, elaborate multicellular organisms, and, finally, a highly intelligent, tool using, sociable, loquacious biped – requires an additional explanatory principle.

Over the years there have been countless *non*-Darwinian theories about this important evolutionary trend. Perhaps the earliest "modern" theorist to advance the idea that there is an inherent tendency in nature toward increased complexity was the eighteenth to nineteenth century naturalist Jean-Baptiste de Lamarck.[2] (I will have much more to say about Lamarck in Chapter 5.) A half century later, the renowned English polymath Herbert Spencer elevated the idea into a "universal law of evolution" that encompassed everything from physics to ethics.[3]

Darwin was strongly opposed to such deterministic theories, needless to say. In *The Origin of Species* he wrote "I believe in no fixed law of development."[4] And again, "I believe...in no necessary law of development."[5] However, Darwin never specifically addressed the evolution of complexity as such, nor even – more notoriously – the origin of species. (Of course, he was fascinated by complex organs, such as the human eye, and suggested how these might have emerged over time from simpler forms.)

Many generations of biology students have learned that Darwin was a convinced gradualist who frequently quoted the popular canon of his day, *natura non facit saltum* – nature does not make leaps. The phrase appears no less than five times in *The Origin of Species.*[6] Although Darwin provided us with many insights about the evolutionary process, he left it to future generations to explain biological complexity.

For this reason, many post-Darwin scholars viewed his theory as insufficient, or at least incomplete, and in the early twentieth century the so-called theory of emergent evolution was advanced as a way of reconciling Darwin's gradualism with the appearance of "qualitative novelties" and, equally important, with the long-term trend in evolution toward new levels of organization and complexity.

Emergent evolution theory had several prominent advocates, but the leading figure in this movement was the comparative psychologist and prolific writer, Conwy Lloyd Morgan.[7] Unfortunately, Lloyd Morgan portrayed the evolutionary process as an unfolding of inherent tendencies, which he ascribed to divine creation. His vision was, of course, rejected by the biologists of his day.

A Minor Dark Age

But far more damaging to the emergent evolution theory was the rise of the science of genetics in the 1920s and 1930s, along with the seminal theoretical work of Ronald Fisher, J.B.S. Haldane, Sewall Wright and others, and the triumph of an analytical, experimental method in biology. During what, in retrospect, was a minor dark age for complexity theory that spanned much of the twentieth century, a reductionist, gene-centered, incremental approach to evolution prevailed. Biologists mainly treated complexity as a non-problem, or an epiphenomenon. It did not require any special explanation. Evolution was all about the machinations of "selfish genes," according the biologist/popularizer Richard Dawkins. (What is commonly known as neo-Darwinism will be discussed in Chapter 5.)

To be sure, there were a few "points of light" during this era. Most notable was the discussion of complexity by J.B.S. Haldane and Julian Huxley in their pioneering 1927 textbook *Animal Biology*.[8] Also significant was the early work on the evolution of complexity by biologists G. Evelyn Hutchinson and John Tyler Bonner.[9] In the latter part of the twentieth century, the biologist Lynn Margulis became a champion for the role of "symbiogenesis" in evolution – especially in relation to the origin of eukaryotic cells but also more broadly.[10] Margulis viewed symbiosis as a major cause of increased complexity in living systems over time.[11]

The Re-emergence of Emergence

The rise of the complexity theory movement toward the end of the twentieth century, inspired by new mathematical modeling tools, gave rise to a new generation of non-Darwinian evolutionary theorists, mainly outside of biology. For instance, the computer scientist and algorithm pioneer John Holland, in his 1998 book *Emergence*, asked: "How do living systems emerge from the laws of physics and chemistry...Can we explain consciousness as an emergent property of certain kinds of physical systems?"[12]

Elsewhere in the book Holland proposed what amounted to the antithesis of the entropy law (the Second Law of Thermodynamics), namely, an inherent tendency of matter to organize itself. Holland illustrated with a metaphor. Chess, he said, is a game in which "a small number of rules or laws can generate surprising complexity."[13] Biological complexity arises from similar (still hidden) rules, he suggested.

The well-known theoretical biologist Stuart Kauffman in his early writings also aspired to discover the underlying laws of evolution, although he too could only offer us a promissory note. In his popular 1995 book, *At Home in the Universe*, Kauffman asserted that "order is not accidental, that vast veins of order lie at hand. Laws of complexity spontaneously generate much of the order of the natural world.... Order, vast and generative, arises naturally."[14] He called it "order for free." In a later book, Kauffman also speculated about a possible "fourth law of thermodynamics," an inherent, energy-driven tendency in nature toward greater diversity and complexity.[15]

There have been innumerable variations on this theme in recent years, with various self-organizing principles being proposed as the engine of complexity.[16] These theories could all be called reductionist in the sense that they posit some underlying force, agency, tendency, or "law" that is said to determine the course of evolution, or at least the evolution of greater complexity, independently of natural selection.

The problem is that these theories explain away the very thing that needs to be explained – namely, the contingent nature of living systems and their fundamentally functional, adaptive properties. The purveyors of these theories often seem oblivious to the inescapable challenges associated with what Darwin called the "struggle for existence" in the natural world, and they discount the economics – the costs and benefits of complexity. Nor can they explain the fact that some 99 percent of all the species that have ever evolved are now extinct. They are re-inventing a wheel that has no spokes.

"A Grand Unified Theory"

Over the course of the past two decades, however, the subject of complexity has finally emerged as a major theme within mainstream evolutionary biology, and a search has been underway for "a Grand Unified Theory" – as biologist Daniel McShea characterizes it – that is consistent with Darwin's great vision.[17]

As it happens, such a theory already exists. It was first proposed in *The Synergism Hypothesis: A Theory of Progressive Evolution* in 1983, and it involves an economic (or perhaps bioeconomic) theory of complexity.[18] Simply stated, cooperative interactions of various kinds, however they may occur, can produce novel combined effects – *synergies* – with functional advantages that may, in turn, become direct causes of natural selection. The focus of the Synergism Hypothesis is on the favorable selection of synergistic "wholes" and the combinations of genes that produce these wholes. The parts (and their genes) that create these synergies may, in effect, become interdependent units of evolutionary change.[19]

In other words, the Synergism Hypothesis is a theory about the unique combined effects produced by the relationships between things. I refer to it as Holistic Darwinism because it's entirely consistent with the natural selection theory, properly understood (see Chapter 2).[20]

Accordingly, it's the functional (economic) benefits associated with various kinds of synergistic effects in any given context that are the underlying cause of cooperative relationships – and of complex organization – in the natural world. The synergy produced by the whole provides the proximate functional payoffs that may differentially favor the survival and reproduction of the parts (and their genes).

Biologist Patrick Bateson illustrates this idea with an analogy. The recipe for a biscuit/cookie is rather like the genome in living systems. It represents a set of instructions for how to make an end-product. A shopper who buys a biscuit/cookie selects the "phenotype" – the end-product, not the recipe. If the recipe survives and the number of cookies multiply over time, it's only because shoppers like the end-product and are willing to purchase more of them.[21]

Although it may seem like backwards logic, the thesis is that functional synergy is the cause of cooperation and complexity in living systems, not the other way around. To repeat, the Synergism Hypothesis is basically an economic theory of emergent complexity, and it applies equally to biological and cultural evolution – most notably in humankind.[22] Indeed, in Chapters 7 and 8 I will propose that social cooperation has been a key to our evolution as a species, and that synergy is the reason why we cooperate. In a very real sense, we invented ourselves.[23]

Synergistic Selection

Biologists John Maynard Smith and Eörs Szathmáry, in their path-breaking work in the 1990s on the "major transitions" in evolution, came to the same conclusion independently about the causal role of synergy – although they graciously acknowledged the priority of my 1983 book in one of their two books on the subject.[24] They applied their version of the Synergism Hypothesis specifically to the problem of explaining the emergence of new levels of biological organization over time.

Maynard Smith also proposed the concept of Synergistic Selection in a 1982 scientific paper as (in effect) a sub-category of natural selection. He illustrated with a formal mathematical model that included a term for "non-additive" benefits – that is, when 2+2=5 (or more). The idea is also distilled in the catchphrase "the whole is greater than the sum of its parts," which traces back to Aristotle in the *Metaphysics*.[25] Synergistic Selection refers to the many contexts in nature where two or more genes/genomes/ individuals have a shared fate; they are functionally interdependent.

The Arc of Evolution

As we shall see, Synergistic Selection is an evolutionary dynamic with a much wider scope than even Maynard Smith envisioned. It includes, among other things, many additive phenomena with combined threshold effects (like the famous straw that broke the camel's back) and, more important, many "qualitative novelties" that cannot even be expressed in quantitative terms.

Synergistic Selection focuses our attention on the causal dynamics and selective outcomes when synergistic phenomena of various kinds arise in the natural world. For it is synergy, and Synergistic Selection, that has driven the evolution of cooperation and complexity over time, including especially the major transitions in evolution. To borrow a famous expression, the arc of evolution bends toward synergy.

The psychologist and evolutionary theorist, Henry Plotkin – in a wide-ranging book about selection processes in evolution – threw down this gauntlet for his colleagues: "A general theory of the biological and social sciences must be able to supply a causal explanation…that encompasses **all** forms of social reality as well as the emergence of living forms…That is the challenge for any theory of selection that claims generality."[26] I believe that the Synergism Hypothesis and Synergistic Selection can meet this theoretical challenge.[27] Life has been a synergistic phenomenon from the get go.

In this book I will argue that it is the unique creative power of synergy that explains what Darwin so eloquently evoked in the summation of his great book, where he spoke about the "grandeur in this view of life, with its several powers, having been originally breathed into a few forms or into one; and that, whilst this planet has gone cycling on according to the fixed law of gravity, from so simple a beginning endless forms most beautiful and most wonderful have been, and are being, evolved."[28]

Looking to the Future

Darwin's vision is inspiring, but we must also look to the future. It's now clear that we are at an evolutionary tipping point – another major transition that Darwin could not have foreseen. The challenge we face was forcefully stated by one of the most distinguished political commentators of the twentieth century, Walter Lippmann, in a 1969 interview just before he died. His words, more than ever, ring true:

> This is not the first time that human affairs have been chaotic and seemed ungovernable. But never before, I think, have the stakes been so high…What is really pressing upon us is that the need to be governed…threatens to exceed man's capacity to govern. This furious multiplication of the masses of mankind coincides with the ever more imminent threat that, because we are so ungoverned, we are polluting and destroying the environment in which the human race must live…. The supreme question before mankind – to which I shall not live to know the answer – is how men will be able to make themselves willing and able to save themselves.[29]

Almost a half century later, Lippmann's "supreme question" remains unanswered. I will focus on this global challenge in the final chapter.

---

[1] Due to space constraints, I will provide only abbreviated endnotes in this book. However, where noted, there will be pointers to "outtakes" at my website: www.complexsystems.org

[2] Lamarck 1914/1809, p. 292.

[3] Spencer 1892/1852, vol. 1, p. 10. (See the outtake at my website.)
[4] Darwin 1968/1859, p. 318.
[5] *Ibid.*, p. 348.
[6] Darwin 1882/1859, p. 133. (See the outtake at my website.)
[7] Lloyd Morgan 1923, 1926, 1933. For a history of emergence theory, see Blitz 1992.
[8] Haldane and Huxley 1927, pp. 234-235. I thank Egbert Leigh for this reference.
[9] See Hutchinson 1965; Bonner 1988; also, Bonner 2003. (See the outtake on this at my website.)
[10] Margulis 1970, 1990, 1993; also, Margulis and Fester, 1991. The term "symbiogenesis" was actually coined by the Russian theorist, Konstantin Mereschkovsky 1909.
[11] Margulis 1998; Margulis and Sagan 2002; also, Carrapiço 2010. Also important was Egbert Leigh's (1977, 1983, 1991, 2010a,b) work on "the common good" and on how group selection can override within-group selection. As Kiers and West (2015) put it: "Symbiotic partnerships are a major source of evolutionary innovation." (For more, see the outtake at my website.)
[12] Holland 1998, p. 2.
[13] *Ibid.*, p. 3.
[14] Kauffman 1995, pp. 8, 25.
[15] Kauffman 2000, p. 5. It should be noted, however, that in his 2008 book Kauffman's views have evolved; he now fully embraces the Darwinian paradigm.
[16] For a brief survey of these theories, see the outtake at my website.
[17] McShea (2015) aspires to find "some single principle or some small set of principles" that could explain the evolutionary trend toward greater complexity. Likewise, biologist Deborah Gordon (2007) laments: "Perhaps there can be a general theory of complex systems, but we don't have one yet." (For more, see the outtake at my website.)
[18] Corning 1983.
[19] The biologist Steven Frank, in a 1997 journal article, made the same point: "Synergism creates associations between [gene] loci, and statistical association may have [selective] consequences similar to physical linkage."
[20] A number of convergent views about the role of cooperation and synergy in evolution will be quoted in various places throughout the book. They have also been assembled in an outtake at my website. For a radical alternative, see Weiss and Buchanan 2009.
[21] Bateson 2013.
[22] A word is in order here about the relationship between synergy and the concept of "emergence". (See the outtake on this at my website.)
[23] For more on this theory, see Corning 1983, 2003, 2005, 2013.
[24] Maynard Smith and Szathmáry 1995, 1999, pp. 22-23. (See the outtake on this at my website.) It should be noted that Klaus Jaffe (2001, 2010, 2016) also independently recognized the role of synergy in evolution.
[25] Maynard Smith 1982a; also 1983. In Aristotle's words, "The whole is something over and above its parts, and not just the sum of them all..." Aristotle, 1961/ca. 350 B.C., Book H, 1045: 8-10.

---

[26] Plotkin 2010, p. 139.

[27] A brief discussion about how this theory relates to the theoretical enterprise known as "synergetics" can be found in an outtake at my website.

[28] Darwin *The Origin of Species*, p. 444. http://darwin-online.org.uk/Variorum/1859/1859-490-c-1860.html (accessed 25 February, 2015).

[29] Quoted in Brandon 1969.

Chapter 2

A New View of Evolution

It seems that the term "cooperation" has become a buzzword in evolutionary biology these days. New books and journal articles on the subject have proliferated. Indeed, one prominent biologist claims that "cooperation is now seen as the primary creative force behind ever greater levels of complexity and organization in all of biology."[1] Another leading theorist calls it "the master architect of evolution."[2]

Yes, but… I will argue here that it's not cooperation per se that has been the creative force, or the architect. Rather, it's the unique combined effects produced by cooperation – the functional synergies – that are the key. Synergies of various kinds have been the underlying cause of cooperation and complexity in evolution. Cooperation may have been the vehicle, but synergy was the driver.

Because the term "synergy" is only vaguely familiar to many of us, it might come as a surprise to learn that synergy is one of the great governing principles of the natural world. It ranks right up there with such heavyweight concepts as gravity, energy, and information in understanding how the world works. Although it travels under many different aliases (mutualism, cooperativity, emergence, density dependence, critical mass, and more), synergy is in fact a ubiquitous phenomenon. It's literally everywhere around us, and it has been a wellspring of creativity and innovation in nature. Most important, it has shaped the overall trajectory of life on Earth, including the rise of humankind – as we shall see. I refer to it as "nature's magic."[3]

What is Synergy?

Synergy can be defined very broadly as the *combined effects produced by the relationships and interactions among various forces, particles, elements, parts, individuals, or groups in any given context* – functional effects that are jointly created and that are not otherwise attainable. The term is derived from the Greek word *synergos*, meaning "working together" or, literally, "co-operating." It's often associated with the expression "the whole is greater than the sum of its parts" (after Aristotle's famous catch phrase), but this is a rather narrow and sometimes even misleading characterization.[4] In fact, synergy comes in many different forms. Sometimes wholes are not greater than the sum of their parts, just different.

One commonplace (frequently-cited) example is water – a versatile liquid composed of the two elemental gases, hydrogen and oxygen. This is an obvious case where the whole has very different properties from its constituent parts. Another example is ordinary table salt. Sodium chloride is a product of two basic elements that are toxic to humans by themselves – the light metal sodium and chlorine gas. But when the two elements are combined they have positively beneficial effects – in small amounts.[5]

Threshold Effects

A very different kind of synergy, called a threshold effect, is illustrated in Pamela Allen's delightful children's story about what happened when all the animals decided to take a boat ride together – *Who Sank the Boat?* It was not the mouse – the last and the smallest of the animals to enter the boat – that was to blame for the disaster.[6] But then you can't blame the elephant either. It was a combined, synergistic effect.

Humans excel at creating novel synergies. Consider one of our most unique inventions, written language. Depending on how the letters are arranged, there can be sixteen possible combinations of the four English letters o, p, s, and t. Many of these combinations mean nothing to us – tpso, ospt, sotp, etc. – but six of them trigger distinct meanings in our minds: stop, pots, tops, spot, post, and opts. It could be said that the four letters "cooperate" to produce different synergistic effects. Even the spatial relationships between the letters can matter. Look at the difference between culdesac and cul de sac. Or consider the difference between recreation and re-creation.

Human technology is also rife with synergy. Take the alloy chrome-nickel-steel, where the three elements together are stronger by about 35% than the sum of the constituents independently and, in the bargain, have rust-free properties. The nickel strengthens the steel and the chrome prevents oxidation.

Another example is a modern automobile, where some 30,000 parts (if you count even the screws) can do things together that would otherwise be impossible if they were simply thrown together in one great heap. (We will explore the role of synergy in the evolution of humankind and human technology in Chapters 7 through 9.)

Synergy in Bacteria

Perhaps the most miraculous synergies of all are found in living organisms. Bacteria, arguably the most plentiful organisms on Earth in terms of their total biomass, are the inventors and masters of a great variety of synergistic effects. Among their many "firsts," we now know that bacteria pioneered in communal living, multicellular organization and even the division of labor, or what should more accurately be called a "combination of labor" (more on this in Chapters 4 and 6).

For example, the process of converting cellulose into useable nutrients inside a cow's rumen is dependent on the combined services of five different strains of symbiotic bacteria. Four of these strains provide different enzymes, each of which accomplishes a specific conversion step, while a fifth strain provides protection for all the others from exposure to oxygen – an ever-present threat to an anaerobic bacterium.[7] Ruminant animals could not digest the grasses they consume without this synergistic partnership.

Another legendary example of synergy in nature can be seen in lichens, the patchy growths that are found on tree limbs, rocks, and even on bare ground in many woodland areas. Lichens are not actually a distinct species. They are a broad category of symbiotic relationships between different species of fungi and various kinds of green algae, or cyanobacteria. There may be 25,000 different lichen varieties altogether. Lichens are renowned for their ability to colonize barren environments, and the key to their success as ecological "pioneers" lies in their complementary talents. The green algae, or cyanobacteria, engage in photosynthesis. They provide energy-capturing services for the partnership, while the fungi have capabilities for surface gripping and water-storage, talents that are especially valuable in harsh environments.

And that's not all. Because each of the two lichen partners may also live independently, symbiosis researcher John Raven was able to do a careful comparison between a local variety of lichens and their free-living cousins nearby. He found that overall nutrient uptake was much better in the lichen partnerships.[8] More recently, it was discovered that many lichens also have a third symbiotic partner, a yeast fungus that enhances their joint performance.[9] So lichens exploit multiple forms of synergy, and these have given them a distinct ecological advantage.

Synergistic Selection and Natural Selection

Accordingly, it is Synergistic Selection, not "classical" natural selection (the individualistic, competitive model), that has defined the overall trajectory of evolution toward greater complexity. The complex systems that have arisen in nature over time are primarily the result of cooperative functional innovations and their combined economic benefits, not atomistic genetic changes alone. If natural selection is all about "selfish genes" (in Richard Dawkins' famous metaphor), Synergistic Selection is all about the joint effects produced by "cooperative genes" and the consequences of these interactions for survival and reproduction.[10] To repeat, Synergistic Selection represents, in effect, a sub-category of natural selection.[11] It's not about cooperation per se but about the unique economic benefits that arise from cooperation.

But we need to begin this story at the beginning. The traditional view, dating back to Darwin himself, is that evolution is driven by a ruthlessly competitive "struggle for existence" (to repeat Darwin's catchphrase), with natural selection as the "decider" – to borrow a term from a former U.S. President. However, this formulation is a caricature – an example of how we can sometimes become the captives of our concepts.

What lies at the core of the theoretical paradigm that assigns a preeminent role in evolution to all-out competition is the very concept of natural selection itself. It began with Darwin, who initiated the practice of treating natural selection as though it were an external selecting agent, or "mechanism" out there in the environment somewhere. (It's well known that the inspiration for Darwin's idea was artificial selection – the techniques used by plant and animal breeders to re-shape a species to their liking.) As Darwin expressed it in *The Origin of Species:*

> Natural selection is daily and hourly scrutinizing throughout the world, every variation, even the slightest; rejecting that which is bad, preserving and adding up all that is good; silently and insensibly working, whenever and wherever the opportunity offers, at the improvement of each organic being in relation to its organic and inorganic conditions of life.[12]

Elsewhere Darwin spoke of how "natural selection acts solely by and for the good of each [organism]..."[13] And again, "natural selection works solely by and for the good of each being..."[14]

Modern-day evolutionists are also prone to personify, or "reify" natural selection, treating it as if it were an active selecting agency, or literally some kind of external force. "By what *force* [his italics] of evolutionary dynamics," asks the distinguished biologist Edward O. Wilson in his popular book *The Social Conquest of Earth*, "did our lineage thread its way through the evolutionary maze?" Wilson's answer: "Natural selection, not design, was the force that threaded this needle."[15]

The problem is that natural selection does not do anything. Nothing is ever actively selected (although predator-prey interactions and sexual selection are qualified exceptions), and it's certainly not literally a force.

Natural selection could be called an "umbrella term" – a broad, open-ended category that applies to whatever factors are responsible in any given context for causing differential survival and reproductive success. It's a way of characterizing the functional consequences – the payoffs – of any change in the relationship between a given organism (or group of organisms) and the environment, inclusive of other organisms. In other words, it's a metaphor, not a distinct causal mechanism.

Proximate versus Ultimate Causes

To use the biologist Ernst Mayr's famous distinction, it is the "proximate" (functional) effects arising from any change in an organism–environment relationship that are the causes of the "ultimate" (evolutionary) changes in any interbreeding population or species over time. The famed Behaviorist psychologist B. F. Skinner aptly called it "selection by consequences."[16] Mayr described evolution as a "two-step, tandem process" – meaning variations of various kinds coupled with "selective retention" of favorable alternatives.[17]

In other words, causation in evolution also runs backwards from our conventional view of things. In evolution, functional effects are also causes. It's not natural selection that "drives" the selective advantages. It's the functional advantages that drive selection. (Recall Patrick Bateson's recipe for biscuits/cookies.) Indeed, in the great flowing river of evolutionary time, almost every effect is also a cause of something else, and every cause arises from some prior effect.

Many things, at many different levels, may be responsible for bringing about changes in an organism–environment relationship and differential survival. It could be a functionally-significant genetic mutation, a chromosome transposition, a change in the physical environment that affects development (ontogeny) or perhaps alters the local food supply, or a change in one species that impacts on another species. Very often it's a change in behavior that results in a new organism–environment relationship. (We'll talk more about how behavior has shaped evolution in Chapter 5.)[18] Thus, genetic changes may be either a cause or a product of the evolutionary process, or both. Biologist Kevin Laland and his colleagues, in a recent paper, refer to this dynamic as "reciprocal causation."[19]

Often, in fact, a whole sequence of changes may ripple through a pattern of relationships. For instance, a climate change might alter the local ecology, which might prompt a behavioral shift to a new habitat, which might encourage a shift in nutritional habits, which might precipitate changes in the interactions among different species, resulting ultimately in the differential survival and reproduction of alternative traits and the DNA coding sequences (the genes) that support them.

One illustration of this dynamic can be found in the long-running research program among "Darwin's finches" in the Galápagos Islands, led by zoologist Peter Grant and his wife and colleague, Rosemary. It's well known that birds often use their beaks as tools, and that their beaks tend to be specialized for whatever food sources are available in any given environment. In the Galápagos Islands, the Grants have observed many changes over time among its fourteen closely-related finch species in response to major environmental fluctuations. During drought periods, for instance, the larger ground finches with bigger beaks survive better than their smaller cousins – surprisingly enough. Small seeds become scarce during the lean years, so the only alternative food source for a seed-eater is the much larger, tougher seeds that must be cracked open to get at their kernels. Birds with bigger, stronger beaks have a functional advantage, and this is the proximate cause of their differential survival.[20]

Behavioral "Selection"

Another example, closer to home (for me), illustrates the role of behavioral "selection" in evolution. In the rainforest of the Olympic National Park, in the state of Washington, there is intense competition among the towering evergreen trees that crowd the forest canopy. The hemlocks produce the most seeds by far and are the best adapted to growing in the competitive, low sunlight conditions of the park. However, it's the Sitka spruce that dominate, and the reason is that the many Roosevelt elk in the park feed heavily on young hemlock trees but do not feed on the Sitka spruce. So, it's the food preferences of the elk that are the proximate cause of differential survival among the hemlock and spruce trees.[21]

In sum, natural selection is not, at heart, a "mechanism" but a dynamic functional process – an "economic" process. It's located in the relationships and interactions among living organisms and their environments, and in the outcomes for survival and reproduction. It's the very opposite of how selection occurs under artificial selection, or in economic markets for that matter, where there is an active selecting agent (a cookie buyer). Thus, paradoxically, natural selection is a concept that both highlights and masks the underlying causal dynamics in evolution. (There are similar problems with how biologists have commonly used the terms "competition" and "cooperation" over the years. We will re-examine these two terms in Chapter 3.)

The Synergism Hypothesis

What John Maynard Smith labeled Synergistic Selection is similar in nature to the broader concept of natural selection in that it identifies an important sub-set of the phenomena that have influenced the course of evolution. From the very origins of life to the latest technological innovations in human societies, synergistic effects of various kinds have been the drivers for the emergence of cooperation and complexity over time (as we shall see). As noted in Chapter 1, this theory is called the Synergism Hypothesis, and it goes back to my 1983 book with that title.[22] The theory was also proposed independently by Maynard Smith and Eörs Szathmáry more than 20 years ago.[23] However, in an era when evolutionary biologists were fixated on the idea that genetic mutations are the primary cause of evolutionary change, the Synergism Hypothesis was mostly ignored.

But times have changed. A major theoretical frame shift occurred in the latter part of the twentieth century when biologists finally began to focus on the problem of explaining the evolution of cooperation and complexity, and when mutualism – symbiosis between different species – also came to be recognized as an important aspect of the natural world.

Nowadays, the organism itself is recognized as a major change agent in evolution (see Chapter 5).[24]

Early on, the effort to explain social cooperation was gene-focused and was widely associated with biologist William Hamilton's inclusive fitness theory, which Maynard Smith dubbed "kin selection."[25] The assumption back then was that cooperation necessarily requires altruism and therefore could not evolve unless the sacrifices made by a cooperator were offset by benefits to close relatives. As the legendary twentieth century biologist J.B.S. Haldane put it in some barroom bravado (actually, a London pub), "I would gladly give up my life for two brothers or eight cousins."[26]

Problems with Kin Selection

Among the many problems with kin selection theory, it turns out, is that the exceptions disprove the rule. "Hamilton's rule," as it is reverently called by his many acolytes, cannot account for the proliferation of mutualistic relationships between different species, at least not without bending the rule into a theoretical pretzel.[27] Nor can it account for the abundance of cooperative behaviors in the bacterial world, where team efforts are ubiquitous, and where promiscuous "horizontal" gene exchanges occur among different organisms rather than by "vertical" transmission only between parents and their offspring.

Most importantly, kin selection is insufficient to explain the emergence of biological complexity over time, including especially what Maynard Smith and Szathmáry characterized as the "major transitions" to new levels of biological organization – from bacteria to complex human societies (see Chapters 6 to 9).

The application of game theory (which should really be called social interaction theory) to evolutionary biology in the 1980s also dispelled the notion that cooperation is necessarily altruistic.[28] Far more common is what could be called egoistic cooperation. Most forms of cooperation involve costs and benefits to each of the parties. It is driven by the economic advantages, and it's most likely to occur when it's mutually beneficial for all parties – win-win, to use the cliché. Nowadays, many evolutionary theorists have embraced what has been termed "multi-level selection theory," the idea that evolution can be influenced at various levels of biological organization, from genes to ecosystems, and that a cooperative partnership, and sometimes an entire social group can become an interdependent unit of evolutionary change.

This is where synergy comes into the picture. The Synergism Hypothesis and Synergistic Selection are now finally emerging from the shadows in evolutionary biology. These concepts tell us where to look for the underlying causes of cooperation and complexity in the natural world. They direct our attention to the combined functional effects that are produced by the relationships between things, or individuals. Kinship, it seems, is neither necessary nor sufficient to explain cooperative phenomena. Synergy, on the other hand, is necessary – although it's obviously not sufficient.[29]

Kin selection is most relevant where altruistic forms of cooperation are involved. Otherwise, social cooperation may have more to do with the proximity of the actors and the economic advantages in cooperating than with genealogy. Indeed, there is much evidence in behavioral biology of cooperation among non-kin and, conversely, innumerable examples of close kin that do not cooperate.[30]

Taking the Measure of Synergy

Many things can influence the likelihood of cooperation in the natural world – the ecological context, specific opportunities, competitive pressures, the risks (and costs) of cheating or parasitism, effective policing, genetic relatedness, biological "pre-adaptations," and especially the distribution of costs and benefits. (The much-debated problems of cheating and free-riding will be discussed in Chapter 3.) However, an essential requisite for cooperation is functional synergy.

Just as natural selection is agnostic about the sources of the functional variations that can influence differential survival and reproduction, so the Synergism Hypothesis is agnostic about how synergistic effects can arise in nature. They could be self-organized; they could be a product of some chance variation; they could arise from a happenstance symbiotic relationship; or they could be the result of a purpose-driven behavioral innovation by some living organism.

It's also important to stress that there are many different kinds of synergy in the natural world, including synergies of scale (when larger numbers provide an otherwise unattainable collective advantage), threshold effects (who sank the boat?), functional complementarities (lichens, for example), augmentation or facilitation (as with catalysts), joint environmental conditioning, risk- and cost-sharing, information-sharing, collective intelligence, animal-tool "symbiosis" and, of course, the many examples of a division (or combination) of labor. (More on the various kinds of synergy in Chapter 4.)[31]

Quantifying Synergy

Synergistic effects can also be measured and quantified in various ways. In the biological world, they are very often related directly to survival and reproduction. Thus, hunting or foraging collaboratively – a behavior found in many insects, birds, fish and mammals – may increase the size of the prey that can be pursued, the likelihood of success in capturing the prey or the collective probability of finding a "food patch." Collective action against potential predators – herding, communal nesting, synchronized reproduction, alarm calling, coordinated defensive measures, and more – may greatly reduce an individual animal's risk of becoming a meal for some other creature.

Likewise, shared defense of food resources – a practice common among social insects, birds, and social carnivores alike – may provide greater food security for all. Cooperation in nest-building, and in the nurturing and protection of the young, may significantly improve the collective odds of reproductive success. Coordinated movement and migration, including the use of formations to increase aerodynamic or hydrodynamic efficiency, may reduce individual energy expenditures and/or aid in navigation. Forming a coalition against competitors may improve the chances of acquiring a mate, or a nest-site, or access to needed resources (such as a watering-hole, a food patch, or potential prey). In all these situations, the synergies are the drivers.

There are also various ways of testing for synergy. One method involves experiments, or "thought experiments" in which a major part is removed from the whole. In many cases (not all), a single deletion, subtraction or omission will be sufficient to eliminate the synergy. I call it "synergy minus one."[32] Take away the heme group from a hemoglobin molecule, or the energy-producing mitochondria from a complex eukaryotic cell, or the all-important choanocytes from sponges (see Chapter 3), or, for that matter, remove a wheel from an automobile. The synergies will vanish.

Another method of testing for synergy derives from the fact that many adaptations, including those that are synergistic, are contingent and context specific, and that virtually all adaptations incur costs as well as benefits. The benefits of any trait must, on balance, outweigh the costs; it must be profitable in terms of its impact on survival and reproduction.

Thus, it may not make sense to form a herd, or a shoal, or a communal nest if there are no threatening predators in the neighborhood, especially if proximity encourages the spread of parasites or concentrates the competition for scarce resources. Nor does it make sense for emperor penguins in the Antarctic to huddle together for warmth at high-noon during the warm summer months, or for Mexican desert spiders to huddle against the threat of dehydration during the wet rainy season. And hunting as a group may not be advantageous if the prey is small and easily caught by an individual hunter without assistance.

Comparative Studies

Another way of testing for synergy involves the use of a standard research methodology in the life sciences and behavioral sciences alike – comparative studies. Often a direct comparison will allow for the precise measurement of a synergistic effect. Some of the many documented examples in the research literature include flatworms that can collectively detoxify a silver colloid solution that would otherwise be fatal to any individual alone; nest construction efficiencies that can be achieved by social wasps compared to individuals; lower predation rates in larger meerkat groups with more sentinels; higher pup survival rates in social groups of sea lions compared to isolated mating pairs; the hunting success of cooperating hyenas in contrast with those that fail to cooperate; the productivity of choanocytes in sponges compared to their very similar, free-swimming relatives called choanoflagellates – not to mention the comparison described earlier between lichen partnerships and their free-living cousins.

A Famous Experiment

A famous experiment in ecology provides an illustration of how the effects of a synergistic combination can be measured and compared to various alternatives. The experiment was designed to study the effects of sunlight and two different fertilizers (nitrate and phosphorus) on the growth of a small woodland flower (*Impatiens parviflora*). One significant finding was that varying amounts of increased sunlight by itself made little difference during the test period. Furthermore, when either nitrate or phosphorous (essential for producing amino acids and proteins) were used alone as fertilizers, it made only an incremental difference. But when the plants were treated with the two fertilizers together, they weighed 50% more at the end of the test period than either of the single-fertilizer groups and almost twice as much as the non-fertilized "controls." The separate contributions of sunlight, nitrogen and phosphorus in plant growth are synergistic – as any skilled gardener already knows.[33]

To appreciate better the role that synergy and Synergistic Selection has played in evolution, there are three other concepts (or, more precisely, three commonplace terms that are sometimes used imprecisely) that we should re-examine before proceeding with an overview of the "evolution story." In the next chapter, we will briefly reconsider the use (and abuse) of the terms "competition" and "cooperation" in biology. In Chapter 4, we will explore the so-called "division of labor" – a prodigious source of synergy in nature – and will then briefly survey some of the many other kinds of cooperation and synergy in the natural world. In Chapter 5, we will critique the "Modern Synthesis" (the neo-Darwinian model of evolution) and describe an emerging new perspective that could be characterized as an "inclusive synthesis." These reassessments will, hopefully, cast the evolutionary process in a different light from the traditional gene-centered model and set the stage for a brief, synergy-guided tour of the history of life on Earth.

---

[1] Michod, 1999, p. xi.

[2] Nowak 2011, p. xviii.

[3] See Corning 2003; also, Corning 1983, 2005, 2013; Corning and Szathmáry 2015.

[4] Aristotle, 1961/ca. 350 B.C., Book H, 1045:8-10.

[5] The scientists Klaus Jaffe and Gerardo Febres (2016) highlight the thermodynamic aspect of synergy. This is a useful approach. (For some critical comments, see the outtake at my website.)

[6] P. Allen 1983.

[7] Van Soest 1994; also, see Wikipedia: http://en.wikipedia.org/wiki/Cellulose (last modified 21 August 2014); see also: http://science.jrank.org/pages/1335/Cellulose-Cellulose-digestion.html#ixzz3CBTQMiuT (accessed 2 September 2014).

[8] Raven 1992.

[9] Spribille *et al.* 2016. (Another remarkable three-way symbiosis – the Azolla, Anabena, bacteria partnership – is described in an outtake at my website.)

[10] See Corning 1996; also, Ridley 2001. (For more on this, see the outtake at my website.)

[11] Synergistic Selection is a concept that is now being utilized with increasing frequency by evolutionary biologists, as reviewed in Corning and Szathmáry 2015. (See also the outtake at my website.)

[12] Darwin 1968/1859, p. 133.

[13] *Ibid.*, p. 229.

[14] *Ibid.*, p. 459. To be sure, Darwin also used a more nuanced phrasing in other places (see Chapter Three), and, in a later edition, he inserted a qualifier before the famous passage quoted above: "It may metaphorically be said…"

[15] E.O. Wilson 2013, pp. 50-51. (Additional examples are cited in an outtake at my website.)

[16] Mayr 1961. Despite some recent nay saying among some biologists, I still find Mayr's distinction to be useful. Among the nay sayers, see especially Laland *et al.* 2011, 2013; Calcott 2013a,b. (For my view of the matter, see the outtake at my website.)

[17] Mayr 1961, 2001. The term "selective retention" was coined by psychologist Donald T. Campbell (1974). (See the outtake on terminology at my website.)

[18] For an in-depth discussion of the role of behavior as a shaping influence in the evolutionary process, see Chapter Five; also, Corning 2014.

[19] Laland *et al.* 2011, 2013.

[20] Grant and Grant 2014; also, Lack 1961/1947; Weiner 1994. (For more on this, see the outtake at my website.)
[21] Warren 2010.
[22] Corning 1983; also, Corning 2003, 2005, 2007a, 2013.
[23] Maynard Smith and Szathmáry 1995, 1999. The term has been used with increasing frequency in evolutionary biology in recent years. See the review in Corning and Szathmáry 2015. (For more on this, see also the outtake at my website.)
[24] See especially Noble 2013; also, Corning 2014.
[25] Hamilton 1964a,b; Maynard Smith 1964.
[26] Maynard Smith (personal communication). See also, Haldane 1932. (For more on Haldane, see the outtake at my website.)
[27] There has recently been a highly technical debate about a "stretched version" of Hamilton's original rule. See especially the insightful analysis by Gintis 2016. (See also the outtake on this at my website.)
[28] Maynard Smith 1982b; also, Binmore 2005.
[29] It seems that others are now converging on this insight. (For a number of examples, see the outtake at my website.)
[30] As noted above, a vexed a debate about inclusive fitness theory has been raging for the past several years. A multi-faceted analysis of synergism by the biologist David Sumpter can be found in his 2010 book *Collective Animal Behavior,* p. 236ff. (For more on this issue, see the outtake at my website.)
[31] Further discussion can be found in Corning 1983, 2003, 2005, 2007a 2013; also, Corning and Szathmáry 2015. (For more on this, see the outtake at my website.)
[32] The term was inspired by "music minus one" – a popular set of recordings (typically a string quartet or quintet) with one instrument omitted, so that a player can fill in with their own part. "Thought experiments" are used quite frequently by scientists as a way of testing ideas. As biologist Richard Dawkins (1982, p. 4) points out: "Thought experiments are not supposed to be realistic. They are supposed to clarify our thinking about reality."
[33] Peace and Grubb 1982.

Chapter 3

How Cooperation Trumps Competition

Most people until quite recently (including many scientists) thought that evolution, and especially Darwin's theory of natural selection, was largely about competition – "nature, red in tooth and claw" as the poet Alfred, Lord Tennyson memorably described it. Call it the high testosterone model of evolution.

For instance, Darwin's friend and vocal champion, biologist Thomas Huxley (famously known as "Darwin's bulldog"), stunned his audience in a public lecture when he characterized evolution as "a gladiator's show," a "relentless combat" where the losers go to the wall.[1]

In the same spirit, the prominent twentieth century evolutionary biologist and notorious curmudgeon (at least theoretically), George C. Williams, wrote a disparaging essay on evolution and ethics entitled "Mother Nature is a Wicked Old Witch."[2]

Not to be outdone, the sometimes salacious popularizer Richard Dawkins wrote with evident relish in *The Selfish Gene*: "I think 'nature red in tooth and claw' sums up our modern understanding of natural selection admirably."[3]

"The War of Nature"

Darwin himself set the tone for these harsh interpretations. "What a book a Devil's chaplain might write on the clumsy, wasteful, blundering low

& horridly cruel works of nature!" Darwin wrote to a colleague in 1856.[4] In *The Origin,* Darwin called it "the war of nature" and repeatedly characterized evolution as "the struggle for existence" (as noted earlier). In fact, Darwin used variations on the term "competition" some 79 times altogether in his text. The word "cooperation" does not appear even once, although Darwin did mention "mutual benefits" a number of times.[5]

More to the point, the very concept of natural selection implies that life is a competitive steeple chase. The famed nineteenth century British social theorist Herbert Spencer called it the "survival of the fittest," and many biologists of that era adopted Spencer's combative term, including Darwin himself in the fifth edition of *The Origin.*[6] (Spencer's slogan was also exploited by the so-called Social Darwinists in the latter nineteenth century to justify rapacious capitalism.)[7] Even a more muted twenty-first century interpretation of Darwinism characterizes evolution as being all about gaining a "selective advantage."

Darwin's dour interpretation of the evolutionary process was partly derived from his own life-long study of the natural world, but it was also strongly influenced by the dark economic vision of the Reverend Thomas Malthus in his famous monograph *An Essay on the Principle of Population* (1798).[8] Malthus foresaw relentless population pressures in humankind, and consequent scarcity, that he believed could only be checked by "the ruthless agencies of hunger and poverty, vice and crime, pestilence and famine, revolution and war."

Darwin Was a Malthusian

Darwin took the Malthusian scenario to heart, as he frankly acknowledged, and applied it to the whole of the natural world. Among the various "laws acting around us," Darwin wrote in *The Origin*, the most important is "a Ratio of [Population] Increase so high as to lead to a Struggle for Life, and as a consequence to Natural Selection..."[9] Malthus's "dismal" prognosis is even embedded in the seldom-used full title of Darwin's masterwork: *On the Origin of Species by Means of Natural Selection, or the Preservation of Favoured Races in the Struggle for Life.*"

Many subsequent generations of evolutionary theorists have adopted a similar worldview.[10] Thus, the contemporary biologist Egbert Leigh tells us that "In nature, organisms compete for resources to live and reproduce… competition drives adaptive change."[11] Leigh even goes so far as to claim that competition "drives the evolution of mutualism" in the natural world.

Well, it's more complicated than that. Competition, like any other influence (say, a climate change), may create a context in which it becomes advantageous to cooperate, but, to repeat, it's the *payoffs* from cooperation (the synergies) that are the chief "cause" of cooperation. As economists Samuel Bowles and Herbert Gintis observe in their book-length study of the subject, the key to cooperation is "a high ratio of benefits to costs…the net benefits."[12]

Semantic Inflation

There's no doubt that competition plays an important role in evolutionary change, but its relative importance has been exaggerated by what could be described as some semantic inflation. The very word "competition" connotes a direct contest between two or more actors in what the game theorists would call a zero-sum relationship; if any one actor gains a benefit, it must come at the expense of another "player."

However, biologists generally recognize two very different forms of competitive interactions. When two animals literally fight over a resource, or a mate, it's referred to as "interference" competition. But there are a great many instances in which the relationships between organisms are more oblique and indirect, say when two adjacent plants both seek to utilize the same limited supply of nitrogen in the soil, or two different predators happen to pursue the same prey. This is referred to as "exploitative" competition, and it can occur without any of the actors even being aware of the challenge. It's not really competition as we normally think of the term.[13] Very often it's simply a matter of which functional variant of a given trait works better or worse in a particular context. (Recall the Galápagos Islands finches and the Olympic rain forest trees described earlier.)

Indeed, there are many mathematical models in population genetics and ecology where "selection coefficients" are utilized to model and predict changes over time in the relative numbers of different traits in a population, but where no functional interaction is assumed to exist between the individuals at all. Although this might be characterized as statistical competition, in fact it's no such thing. A species may arise, thrive for a time, and go extinct for reasons that have nothing to do with the actions of some other "competitor" species. At other times, various species may engage in niche partitioning – in effect avoiding competition with each other by specializing on different resources, although their relative numbers may still change over time as their habitats change. Still other species find ways to regulate their numbers and avoid the Malthusian trap, while many more cooperate, as we shall see. Indeed, an animal can win the lottery simply by producing more offspring.[14] By the same token, there are species that have thrived for many millions of years without any major change. Fossilized ancestors of a modern-day deep-sea bacterium were recently discovered near Australia that were two billion years old.[15]

With all due respect for the competitive dimension of evolution, we need a more balanced view of the process, and, in fact, a more dualistic paradigm is now becoming more widely accepted in biology.

The Collective Survival Enterprise

It might be helpful to pause briefly and redefine the basic problem. For starters, we should de-escalate Darwin's rhetoric and call it the "challenge of existence." The basic, continuing, inescapable problem for every living organism is survival and reproduction, and the outcome is always contingent. Life is quintessentially a "survival enterprise." It is dependent on the ability of an organism to meet an array of ongoing, context specific basic needs and to reproduce successfully. This is the fundamental vocation, and the enduring preoccupation, of all living things (including even humankind, in case you hadn't noticed).

Sometimes the survival enterprise may entail competition, but at other times a species may benefit from cooperation; cooperation and competition are not mutually exclusive survival strategies. And when a cooperative relationship materially affects the survival chances of the cooperators – when there is a functional interdependency – then it can legitimately be characterized as a "collective survival enterprise." Indeed, cooperation is often exploited in nature as a way to achieve a competitive advantage. You could call it competition via cooperation. Alternatively, cooperation might allow a species to create a wholly new niche and circumvent direct competition, like the many examples of symbiosis.[16]

To repeat, it's cooperation (or, better said, the synergies produced by cooperation) that has been primarily responsible for the progressive evolution of ever more complex species on Earth, culminating in humankind (as I will argue). Perhaps this is the reason why many biologists these days prefer to define natural selection in less sanguinary terms simply as "differential survival and reproduction."

In fact, Darwin himself provided a more balanced and nuanced rendering of the evolutionary process elsewhere in *The Origin*. One of the most widely quoted passages occurs in his final recapitulation:

> It is interesting to contemplate an entangled bank, clothed with many plants of many kinds, with birds singing in the bushes, with various insects flitting about, and with worms crawling through the damp earth, and to reflect that these elaborately constructed forms, so different from each other, and dependent on each other in so complex a manner, have all been produced by laws acting all around us…[17]

In another passage, Darwin also qualified what he meant by the "struggle for existence":

> I should premise that I use the term Struggle for Existence in a large and metaphorical sense, including dependence of one being on another, and including (which is more important) not only the life of the individual, but success in leaving progeny…Two canine animals in a time of dearth, may truly be said to struggle with each other over which shall get food and

live. But a plant on the edge of a desert is said to struggle for life against the drought, though more properly it should be said to be dependent on the moisture...The mistletoe is dependent on the apple and a few other trees, but can only in a far-fetched sense be said to struggle with these trees...In these several senses, which pass into each other, I use for convenience sake the general term of struggle for existence."[18]

Defining Cooperation

If "competition" is a term that evolutionary theorists have sometimes used a bit too expansively, the opposite has been true with respect to the term "cooperation." Until quite recently, cooperation was, for various reasons, often defined very narrowly and its scope was assumed to be quite limited. For one thing, it was widely believed that cooperation did not include mutualism (symbiosis) between members of different species and that mutualism was not, in any case, of much importance. Also excluded, for the most part, were collective behaviors that involve a synergy of scale and "public goods" that are equally shared by all participants – behaviors such as herding, flocking, mobbing, and huddling.

Evolutionary theorists often confined the term cooperation to direct, purposeful interactions between two or more members of the same species under conditions that were assumed to be asymmetrical. That is, cooperation must entail "altruistic" costs for a "donor" in return for benefits to a "recipient" (as noted earlier). In effect, cooperation was put into a theoretical strait jacket. Even today there remain some holdouts.[19]

To make matters worse (theoretically), the introduction of game theory into evolutionary biology by John Maynard Smith in the 1980s highlighted what came to be viewed as a serious obstacle to cooperation, namely, the presumed tendency of a cooperator to always cheat, or to become a "free-rider" that would exploit and harm other cooperators if given an opportunity.[20] This theoretical conundrum is commonly known as the Prisoner's Dilemma, and it reflects the presumed bias of natural selection toward the relentless pursuit of competitive self-interest. Over time, the Prisoner's Dilemma inspired a great variety of theoretical workarounds, with many of the proposed models ranging far beyond the original, narrowly conceived, hypothetical, and (in retrospect) highly improbable social dilemma.[21]

After almost two decades of arm-chair theorizing (or rather, computerized theorizing) by numerous biologists, and with precious little field research to test the plethora of theoretical models that were proposed, the game theory pioneer himself (Maynard Smith) and his co-author Eörs Szathmáry in their 1995 book on *The Major Transitions in Evolution* concluded that "The intellectual fascination of the Prisoner's Dilemma game may have misled us to overestimate its evolutionary importance."[22]

Loading the Dice

One of the leading game theorists of the past generation, the mathematician Ken Binmore, is even more critical. He believes that the progress of social theory was seriously retarded by the fact that the only formulation developed by early theorists to model cooperation was the Prisoner's Dilemma.[23] As Binmore observes: "A whole generation of scholars swallowed the line that the Prisoner's Dilemma embodies the essence of the problem of human cooperation…On the contrary, it represents a situation in which the dice are as loaded against the emergence of cooperation as they could be…Rational players don't cooperate in the Prisoner's Dilemma because the conditions necessary for rational cooperation are absent." It's rather like explaining why people drown when they are thrown into Lake Michigan with their feet encased in concrete, Binmore concluded.[24]

In order to appreciate better the role that cooperation has played in evolution, we should start by clarifying what the term means. Binmore, in his writings, take a minimalist approach. Cooperation exists, he says, whenever two or more individuals in some way coordinate their behavior – say, when automobile drivers obey the rules of the road, or when pedestrians act to avoid one another on sidewalks.[25]

I prefer a more functionally-oriented definition: Cooperation exists when two or more elements, genes, parts, individuals, or groups interact in such a way that they produce combined effects – synergies. Mere accommodation or avoidance is not enough. One well-documented example of cooperation at the molecular level is the so-called *Hox* gene cluster, a remarkably conserved set of genetic coding sequences, with variations in number from 8 to more than 40, that collectively determine the basic body plans of *Drosophila* flies, human embryos, and everything in between.[26] Another example, at the "macro" level, is the coordinated actions of thousands of honeybee foragers, aided by their famous "waggle dance," which enables the colony as a whole to gather enough pollen to survive the winter months and reproduce the next generation of honeybees. In both of these examples of cooperation, there are combined, synergistic outcomes.

Pathways to Cooperation

One approach to understanding how cooperation – and synergy – can evolve in the natural world is "structural." It's focused on identifying what are likely to be successful contexts and patterns of interaction among the various participants. The mathematical biologist Martin Nowak has identified five different "pathways" to cooperation, as he calls them: Kin selection, direct reciprocity, indirect reciprocity, network reciprocity, and group selection.[27]

Nowak believes that kin selection plays only a limited role. Direct reciprocity, on the other hand, involves a quid pro quo, or tit-for-tat. The third category, indirect reciprocity, occurs when you scratch a friend's back, then he/she does likewise for a third party, and so on, until someone eventually does the same for you. It's enshrined in the old saying, "what goes around comes around."

In contrast, network reciprocity envisions a structured population that might allow cooperators, cheaters, and defectors to coexist but where cooperators will dominate. To borrow a metaphor from Maynard Smith, it could be called a "primordial pizza." The mix of ingredients may vary, but some combinations taste better than others.[28]

The final category, group selection, can be favorable to cooperation if it enables a harmonious group to prevail over other groups that are dominated by defectors and non-cooperators. (This category, as we shall see in Chapter 7, was identified by Darwin himself as a key factor in human evolution, and many modern theorists have belatedly come around to his view, after several decades of rejecting it.)[29]

Two other structural arrangements may also be important in dealing with the very real risks of cheating or free riding in cooperative relationships. One is a relationship that approximates what is known as a Nash equilibrium – a concept developed by the Nobel prize-winning mathematician John Nash (whose mental illness and troubled life was reenacted in the movie *A Beautiful Mind*). In essence, a Nash equilibrium refers to a relationship in which there is a stable, self-reinforcing kind of cooperation. This can occur when all the participants are benefiting as much as they can (optimizing) and when no further gains are possible for anyone by changing their strategies, or "defecting."

Sculling versus Rowing

The other structural category of cooperative relationships, one that is often underrated or even overlooked, involves the many contexts in which there is a functional interdependency – when the participants depend on the benefits (the synergies) in such a way that cheating or free-riding would undermine those benefits and become self-defeating. Whenever such interdependencies exist, and they are more common than is often appreciated, a cooperative relationship is, in effect, self-policing. These are examples of what Maynard Smith termed an evolutionarily stable strategy (or ESS).

A hypothetical example was provided by Maynard Smith and Szathmáry in *The Major Transitions in Evolution.* It involves a game theory model where two oarsmen in a rowboat are seeking a common objective, say crossing a river. If the two oarsmen are using a sculling arrangement, each one has a pair of oars and they row in tandem. In this situation, it is easy (at least in theory) for one of the oarsmen to slack off and let the other one do the hard work. This corresponds to the classical two-person game. However, in a two-person rowing model, each oarsman has only one opposing oar. Now their relationship to the performance of the boat is interdependent; if one of the oarsmen slacks off, the boat will go in circles.

A biological example of the rowing model – among many others – can be found in chromosomes. If a selfish gene, in Dawkins' metaphor, defects (and there are numerous well-known cases of deleterious mutant genes), it undermines the ability of the genome to reproduce the organism successfully.[30] Many symbiotic mutualisms also fit the rowing model – such as the obligate relationship between a ruminant animal and its cellulose digesting gut bacteria.

It should also be stressed that there is now a large body of research regarding the role of policing and punishments as enforcers of cooperative relationships in the natural world. As biologists Timothy Clutton-Brock and Geoff Parker conclude in a review article on the subject: "In social animals, retaliatory aggression is common. Individuals often punish other group members that infringe their interests, and punishments can cause subordinates to desist from behavior likely to reduce the fitness of dominant animals. Punishing strategies are used to establish and maintain dominance relationships, to discourage parasites and cheats, to discipline offspring or prospective sexual partners and to maintain cooperative behavior."[31] Policing behaviors have been documented in, among others, social insects,[32] naked mole-rats,[33] primates,[34] and, needless to say, *Homo sapiens*.

Especially relevant for understanding the evolution of cooperation in humankind is the theory and research related to what has been termed "strong reciprocity theory," a formulation in which cooperative behaviors are backed by aggressive sanctions to control cheating, including even altruistic punishments.[35] Call it tit-for-tat with teeth.

The Corporate Goods Model

A second broad approach to understanding the evolution of cooperation involves a frame shift to focus on the *functional* consequences, or the costs and benefits for each of the participants – in other words, the synergies and how they are distributed. While structural factors may be important as facilitators or constraints, in fact it is the benefits derived from cooperation, net of the costs (the profits) that are the primary cause of cooperation at all levels.[36] To repeat, cooperation may be the vehicle, but synergy is the driver.

A functional (economic) approach to cooperation is embedded in what I call the Corporate Goods model. Corporate goods are benefits that are jointly produced by two or more participants. However, unlike public goods that are indivisible and always equally shared, corporate goods can be divided up in different ways among the actors – for example, the distribution of meat from a large game animal procured by a group of hunters, or the sales receipts for a large industrial organization like Walmart. In order for a corporate goods relationship to be favorably selected (Synergistic Selection) and be "sustainable" (or evolutionarily stable), the following conditions apply:

i. There must be an overall "profit" (the benefits must outweigh the costs).
ii. The benefits to each participant (direct, indirect, or both) net of the costs must be positive.
iii. The relationship is supported/enforced by one or more of the following:
    a. There is a functional interdependence, so that the relationship is self-enforcing (as in the "rowing model");
    b. There is no better alternative (i.e., a more favorable benefit-cost ratio) available to any participant by defecting to some other relationship (a Nash equilibrium);
    c. The benefits may be reduced or denied to any defector;
    d. There is some other punishment/sanction for defecting (e.g., ostracism, or denial of other benefits in a multifaceted co-operative relationship, etc.).

The Corporate Goods model differs from the classic game theory paradigm in several ways. First, it allows for the payoff matrix to be manipulated to "incentivize" the players and perhaps adjust the payoffs over time as conditions require. Second, it entails one or more measures to ensure that the promised payoffs are in fact delivered. And third, it can be structured as an ongoing game, with a replenishment of the benefits to ensure continuity in the relationship over time.

It is not, therefore, a substitute for the plethora of one-shot, bilateral transactions in social life but a model for a different kind of cooperative relationship. Not coincidentally, it also happens to resemble a ubiquitous form of cooperation in human societies which, I believe, has been neglected in game theory models. (Of course, corporate goods can also be achieved with coercion – for example in a slave system, or when a modern-day worker has no alternative to an onerous, low-wage job.)

It's also important to note that some theorists tend to overstate the cost side of a cooperative relationship. In reality, cooperation often involves what the economists refer to as "opportunity costs" or "marginal costs." If an animal must search for food, the opportunity cost of hunting in a group is only the cost difference compared with hunting alone. In fact, it could be that group hunting reduces the average cost to each participant and increases the probability of success. And if a group of hunters can capture much larger prey, the marginal cost of sharing it with others may be insignificant or non-existent. There could be an unalloyed positive synergy.

Cooperation and Synergistic Selection

Many years ago, the distinguished novelist and polymath Arthur Koestler observed that "true novelty occurs when things are put together for the first time that had been separate."[37] He was talking about synergy, of course. Cooperation can produce novel combined effects – synergies that yield immediate benefits for one, or all, of the cooperators. If this cooperative relationship endures and perhaps spreads, it would be an example of Synergistic Selection at the proximate behavioral level. And if these novel relationships provide a selective advantage in relation to the overall survival enterprise, it may result in Synergistic Selection at the ultimate level as well. To repeat, it becomes, in effect, a sub-category of natural selection.

Nowadays, with the theoretical underbrush having been (largely) cleared away from evolutionary biology, it's evident that cooperation is not a problem after all. Our theoretical toolkit is fully capable of dealing with the reality out there in the natural world. Cooperative phenomena are ubiquitous in nature – from bacterial biofilms to the pervasive symbiotic partnerships between species at all levels, social organization in numerous animal species, and, not least, the emergent complexity and interdependence of ecological communities like reef corals and tropical rain forests (as we shall see).[38]

Furthermore, cooperation is perfectly consistent with natural selection – properly understood. It is evident that competition and cooperation have both played important parts in the "epic of evolution" (in Edward Wilson's term) and that the selfish gene is also a cooperative gene. A bioeconomic approach to cooperation, and a shift of focus to the creative role of various kinds of functional synergy in evolution, helps us to understand why. In the next chapter, we will examine in more detail some of the many different kinds of synergy in the natural world.

---

[1] From Huxley's notorious Romanes lecture, "Evolution and Ethics," in 1893. Reprinted in Nitecki and Nitecki eds. 1993.

[2] Williams 1993.

[3] Dawkins 1989/1976, p. 2.

[4] From a letter to J.D. Hooker, 13 July 1856. http://www.darwinproject.ac.uk/letter/entry- 1924 (accessed 17 February, 2015).

[5] Darwin 1882/1859 (Sixth Edition). In *The Descent of Man* (1874/1871) Darwin's emphasis shifted. Here he used the term "social" more than 100 times, the term "mutual" 32 times, "co-operation" four times, and "competition" only 11 times.

[6] Spencer http://en.wikipedia.org/wiki/Survival_of_the_fittest (last modified 25 February, 2015).

[7] See for example Carnegie 1992/1889. (For a quote, see the outtake at my website.)

[8] Malthus 2008/1798, Ch. 2, 18. (Quoted in an outtake at my website.)

[9] Darwin 1968/1859, p. 459.

[10] For details on this history, see Corning 1983. (For more on this, see also the outtake at my website.)

[11] Leigh 2010a.

[12] Bowles and Gintis 2011, p. 197.

[13] For short summaries and discussions of this subject see especially https://en.wikipedia.org/wiki/Competition_(biology) (last modified 1 June 2015); also,

https://en.wikipedia.org/wiki/Competition (last modified 11 June 2015); and https://en.wikipedia.org/wiki/Intraspecific_competition (last modified 9 April 2015). Our perspective on the role of competition also continues to shift. Recent doubts about its importance in producing species diversity are discussed in Singer 2014.

[14] Indeed, much competition is an indirect result of the capacity for living organisms to expand their numbers over time and exploit the available space and resources. These days many theorists see direct competition as less significant as a driver of evolution than ecological and other influences. See for example Sahney *et al.* 2010.

[15] Schopf *et al.* 2015. The authors of this paper describe it as "extreme evolutionary stasis."

[16] See especially Carrapiço 2010; Pereira *et al.* 2012.

[17] Darwin 1968/1859, p. 459.

[18] *Ibid.*, p. 116.

[19] For recent examples of this constricted paradigm, see Sachs *et al.* 2004; Lehmann and Keller 2006; West *et al.* 2011. (See also the outtake at my website.)

[20] Maynard Smith 1982b, 1998.

[21] A brief history of this paradigm can be found in Binmore 2005; also, Nowak 2011. Nowak provides a detailed discussion of the many different "solutions" to the Prisoner's Dilemma. See also https://en.wikipedia.org/wiki/Prisoner%27s_dilemma (last modified 19 February 2016). (For more on this, see the outtake at my website.)

[22] Maynard Smith and Szathmáry 1995, p. 261.

[23] Less well known is the work on a more positive alternative model known as the stag hunt. The basic idea is that if two hunters collaborate to hunt for a stag they can do better than if each one hunts for hare on his own -- synergy! See Skyrms 2004.

[24] Binmore 2005, p. 63.

[25] *Ibid.*

[26] Hunt 1998.

[27] Nowak 2006; also, Nowak 2011.

[28] The term "primordial pizza" has also been used to characterize one of the alternative theories about the origin of life, namely, that it originated on land, perhaps in a layered form.

[29] Darwin 1874/1871. See also Smaldino 2014.

[30] See Crow 1979. Weiss and Buchanan (2009) use the metaphor of a lock and key, where a single defect can render it inoperative. Call it "synergy minus one."

[31] Clutton-Brock and Parker 1995, p. 209.

[32] Ratnieks and Visscher 1989.

[33] Sherman *et al.* eds. 1991, 1992.

[34] de Waal 1982, 1996.

[35] See especially Gintis 2000a,b; Gintis *et al.* 2003; Fehr and Gächter 2000a,b; Falk *et al.* 2001; Sethi and Somanathan 2001; Fehr *et al.* 2002; Fletcher and Zwick 2004; Bowles and Gintis 2011.

---

[36] See also the discussion in Bowles and Gintis 2011; also, Sterelny *et al*. 2013.
[37] Koestler 1967.
[38] A wide-ranging and very readable discussion of the role of cooperation in evolution, with a special focus on the molecular and cellular levels in living systems, can be found in Weiss and Buchanan (2009). They note, for example, that almost every trait embodied in the genome of a complex organism involves a system of genes linked by a network of communication "signals" (pp. 223-225).

Chapter 4

Evolution as a "Combination of Labor"

Social cooperation in nature could more accurately be termed a combination of labor. It involves a purposeful interaction directed toward some joint outcome that may, in turn, exert an influence on the evolutionary process via Synergistic Selection. We can, perhaps, gain a better understanding of the central role of cooperation and synergy in evolution by taking a brief excursion to look at some of its many different varieties. Let's start with what is arguably the most important category of all.

The familiar term "division of labor" has acquired godlike status in economics and, increasingly, in evolutionary biology as well. It was the pioneer economist Adam Smith, in *The Wealth of Nations*, who established the concept as a centerpiece of economic science with his insight that economic progress is fueled by increases in productivity through a division of labor. Smith wrote:

> The greatest improvement in the productive powers of labor, and the greater part of the skill, dexterity, and judgment with which it is anywhere directed or applied, seem to have been the effects of the division of labor .... It is the great multiplication of the productions of all the different arts, in consequence of the division of labor, which occasions, in a well governed

> society, that universal opulence which extends itself to the lowest ranks of the people.[1]

Modern economists wholeheartedly agree. For instance, economist George Stigler, in a now-classic journal article back in the 1950s, characterized the division of labor as "a fundamental principle of economic organization."[2] Over the years there have been innumerable scholarly studies of the division of labor, and introductory economics textbooks these days routinely genuflect to this sacred concept.

Emile Durkheim, one of the founding fathers of modern sociology, went much further. In his seminal treatise on *The Division of Labour in Society* in 1893, Durkheim argued that this principle applied to "all biological organisms generally," and he perceived an important causal relationship between "the functional specialization of an organism" and "the extent of that organism's evolutionary development." He believed that a division of labor was "contemporaneous with the origins of life itself." He called it a "natural law."[3]

A Hot Concept in Biology

Modern biologists seem to have joined the chorus. John Tyler Bonner, a long-time student of complexity in evolution, notes that there is a strong correlation between size and the degree of functional differentiation in the natural world.[4] Carl Anderson and Nigel Franks, in an article on "teams" in animal societies, conclude that a division of labor is much more common than previously thought.[5] Carl Simpson reviews how a division of labor has evolved over time and asserts that "We generally think the benefits…are profound, allowing an organism to simultaneously accomplish several physiological processes."[6] And in a suggestive experiment in yeast, William Ratcliff and his co-workers discovered what seemed to be a built-in readiness to evolve a multi-cellular division of labor.[7]

Bert Hölldobler and Edward Wilson, in their magisterial 2009 book on insect societies with the evocative title, *The Superorganism,* found evidence of a similar tendency in insects. They defined a eusocial insect society as "a colony of individuals self-organized by division of labor and united by a closed system of communication."[8] Wilson expanded on this idea in his popular book on *The Social Conquest of Earth* and concluded that the great advantage enjoyed by eusocial insect species is that they can exploit a division of labor.[9]

It so happens that the underlying concept is hardly a recent discovery. In fact, it was the ancient Greek philosopher Plato who first singled out the central importance of a division of labor in human societies (although he didn't use this term). In his classic dialogue, the *Republic*, Plato wrote:

> No human being is self-sufficient, and all of us have many wants.... [So] let us construct a city beginning with its origins, keeping in mind that the origin of every real city is human necessity.... [However], we are not all alike. There is a diversity of talents among men; consequently, one man is best suited to one particular occupation and another to another.... We can conclude, then, that production in our city will be more abundant and the products more easily produced and of better quality if each does the work nature [and society] has equipped him to do, at the appropriate time, and is not required to spend time on other occupations....[10]

Many other theorists over the past two thousand years have also extolled the seemingly magical influence of a division of labor, including Xenophon, Ibn Khaldun, Bernard de Mandeville, David Hume, Herbert Spencer, David Ricardo, and Alexis de Tocqueville, to name just a few. Perhaps most noteworthy was the great nineteenth and early twentieth century English economist Alfred Marshall. In his pioneering textbook *Principles of Economics* in 1890, he provided a detailed and insightful analysis of how the division of labor works in an industrial society. The benefits are obvious, he said. Dividing up a complex task into simpler pieces allows managers to increase the workers' skill and efficiency while, at the same time, using less skilled workers who can be paid lower wages. During the preceding 70 years, Marshall reported, the productivity of labor in weaving had increased twelve-fold and in spinning by six-fold, on top of a two hundred-fold increase in the 70 years prior to that.

Marshall also asserted that machines should be viewed as an integral part of the division of labor, with the added advantage that they enable managers to replace repetitive "muscle work." Operating and tending machines involves "higher faculties" of intelligence and skill, he wrote. For example, a woman worker in a clothing factory could manage several looms at once, rather than being tethered to a single task. Marshall viewed machines as an unequivocal good because, he claimed, they eliminated monotonous work, the "chief evil" of the factory system. (Of course, we now appreciate that this very much depends on the context; machines can also impose monotonous repetition.) Marshall also acknowledged that machines can both create jobs and eliminate them, but he believed the benefits far outweighed the costs.

It's Really About Cooperation

I prefer to characterize this all-important category of synergistic effects as a combination of labor, first, because it's really about social cooperation and, second, because it's part of a much broader category of cooperative relationships, including many kinds that do not involve dividing up a task into different parts (or "task partitioning," as it is sometimes called). Consider, for instance, a tug-of-war, when everyone is pulling together in the same direction to achieve a collective outcome. Or look at mobbing behaviors – say, when a group of jointly nesting birds gang up to drive off a predator. (Recall also the animals in Pamela Allen's *Who Sank the Boat?*) These, too, are cooperatively-produced, synergistic effects.

What is most important about the division of labor is not simply that some task is divided into smaller sub-tasks, but rather that a number of individuals closely coordinate their efforts to produce a combined result – a synergistic effect that is advantageous in some way – faster, more efficient, less costly, or simply impossible to achieve otherwise.

The textbook example (literally) was provided by Adam Smith in *The Wealth of Nations*.[11] Smith reported that he had personally observed a pin factory where ten workers performing ten different tasks were able to manufacture about 48,000 pins per day. But if each of the laborers were to work alone, attempting to perform all the tasks required to make pins rather than working cooperatively, Smith doubted if, on any given day, they could produce even one or at most 20 pins per man.

The writers of modern-day economics textbooks are fond of using Adam Smith's pin factory as an illustration of the division of labor, but his example downplays the synergy. Another way of looking at the pin factory is in terms of how various specialized skills, tools and production operations were combined into an organized system. The parts of the system included not only the roles played by each of the workers, which had to be carefully coordinated, but also the appropriate machinery, power to run the machinery (and tools to maintain the machines), sources of raw materials, a supportive transportation system and, of course, markets where the pins could be sold to recover the costs.

In addition, the pin factory required management – one or more persons who were responsible for hiring and training workers, for planning the operation, for production decisions, for selling the pins, for payroll and bookkeeping, and so forth. In other words, the economic benefits (the synergies) realized by the pin factory were the result of the total system, including a complex network of production tasks and cooperative relationships.

How do we know it was synergistic? Just imagine what might happen if one of the key pin-making machines broke down, or if its highly trained operator called in sick. The synergy would grind to a halt. Again, I refer to it as synergy minus one. (Another classic example of a combination of labor – Henry Ford's automobile assembly line – is described in an outtake at my website – www.complexsystems.org.)

The Combination of Labor in Sponges

One illustration, among many, of a division (combination) of labor in the natural world can be found in sponges.[12] Although sponges come in many different sizes and shapes, they often look more like an urn or a vase than a typical kitchen sponge. Sponges are also the most primitive of all animals in terms of complexity. In fact, they have sometimes been confused with plants because they are immobile and have no internal organs, no mouth, no gut, no sensory apparatus, nor even a nervous system. They are more like a colony of cooperating independent cells. (There are an estimated 10,000 species of sponges altogether.)

Sponges also earn their living in a very simple way, as filter feeders. They pull water into an internal cavity through large pores in their skin, which consists of an outer layer of surface cells and a gelatinous inner layer supported by a skeleton of thin, bone-like "spicules." The internal cavity of the sponge is in turn lined with a layer of specialized cells called choanocytes that are equipped with whip-like flagella and numerous filaments. These so-called collar cells combine forces to move the water through the sponge and then push it out through a large opening at the top called the osculum. As the water passes through the sponge, the collar cell filaments extract oxygen and food particles, such as microbes and organic debris of various kinds. These nutrients are then distributed throughout the sponge via another specialized set of transporter cells called amoebocytes. The amoebocytes are also responsible for carrying wastes and for manufacturing and distributing various kinds of skeletal materials.

Reproduction in Sponges

Reproduction in sponges is also typically a cooperative effort. Although the freshwater forms frequently reproduce asexually (often by casting off something that resembles seed pods), most sponges are hermaphrodites, meaning that they produce both sperm cells and eggs. The sperm cells are launched into the sponge's cavity and then ejected through the osculum in the hope that they will find their way to another sponge. If a sperm is lucky enough to enter a host sponge, it may be captured by one of its collar cells and then transferred to an amoebocyte, which in turn carries it to an awaiting egg. Eventually, the fertilized egg will become a free-swimming larva and will venture out on its own to find an appropriate site for developing into a new adult. It's a very unusual reproductive system.

That's about all there is to say about the combination of labor in sponges, except for the chemicals they produce to repel potential predators. Their "system" involves fewer cell-types (six) than the number of workers in Adam Smith's pin factory – namely, epithelial cells, pore cells, collar cells (choanocytes), amoebocytes, and two kinds of sex cells. (Some larger sponges also have specialized cells that aid in opening and closing their oscula.)

The point here is that even the minimal level of differentiation and complexity found in sponges is tied directly to the functional effects that the parts produce together – the synergies. Each part is specialized for the role it plays in the system. Each part is also dependent upon the other parts; no part could exist without the services of the others, and only together can they survive and reproduce successfully. Furthermore, the function of each part cannot be understood without reference to its role in the operation of the system as a whole. Nor can we understand the whole without an appreciation of how the parts work together. It's truly a synergistic combination of labor, and a "collective survival enterprise."

The Paradox of Dependency

There are, of course, many different ways in which labor can be divided up and combined in living systems. It could be done according to sex (reproduction is an obvious example), or by age, or in relation to the participants' skills and experience, or sequentially (as in an assembly line), or spatially (as in the manufacture of different automobile parts in different locations), or even by different levels of organization in a hierarchical system – say, families, business firms, trade associations and multiple levels of government. Adam Smith himself described some of the different kinds of cooperative labor, and he was the first economist to suggest that there could even be some partitioning of labor among various nations in relation to what he called a "comparative advantage."[13] Our twenty-first century global economy is rife with examples of the latter, from the banana plantations in the Caribbean to the rubber plantations in Malaysia.

However, there is a deep paradox involved in dividing up and sharing the elements of a job. It creates an interdependency; everyone must do their part or the desired outcome will not be achieved. Imagine (again) what would happen if one of the pin making machines in Adam Smith's pin factory broke down. Or imagine if a sponge were to be deprived of its choanocytes.[14] Or, for that matter, imagine the consequences in Maynard Smith and Szathmáry's rowing model if one of the oarsmen defects. In other words, a "system" can create a built-in enforcer for cooperation. What could be called the paradox of dependency is an underrated aspect of a division of labor. The paradox arises from the fact that, as the value of the system increases for the participants, so does the cost of losing it.

Other Combinations of Labor

In the next few chapters we will explore further the role that the division/combination of labor has played in achieving new levels of biological organization, but it is important to note that this is just one of the many different categories of cooperation and synergy in the natural world. Ten of these categories are briefly described below; they are discussed in more detail elsewhere.[15] (It should be noted that some of these categories overlap – simply because the same phenomenon can often be viewed from different perspectives, or through different prisms. Consider, for example, the different ways of viewing Adam Smith's pin factory – as a division of labor, or as a combination of labor.)

- **A Synergy of Scale** – As mentioned earlier, this form of cooperation involves situations where a number of similar parts, or individuals, act together in a concerted way to accomplish a common outcome. Here, size makes all the difference. A bigger molecule, a bigger organism, a bigger group, or a bigger organization (or nation) may be able to do things that smaller ones cannot. And when it comes to direct competition between groups, the odds are that the bigger unit will prevail.[16] Economists often use the term "economies of scale," or "returns to scale" but many of the synergies produced by being larger, or having a larger number of actors, are not necessarily more economical or cost-efficient but simply more potent. (We will return to this category in relation to human evolution in Chapters 7 through 9.)

Thus, jointly nesting birds may combine forces to mob and collectively drive off a predator. They succeed in doing together what none of them could do alone. Similarly, large colonies of the predatory bacterium *Myxococcus xanthus* are jointly able to engulf and subdue much larger prey than any one or a few of them could do by itself.[17] A large coalition of male lions is generally able to take over and hold a pride of females against smaller groups, or individual males. And in the largest sponges – some taller than a human – the internal walls are elaborately folded, which greatly increases the surface area available for filtering the passing water to meet the increased nutritional needs of a larger organism.

A very different synergy of scale can be seen in the collective behavior of a tropical reef fish called the wrasse. These small fish prey on the abundant supply of eggs produced by the much larger sergeant-major damselfish. A very small group of wrasse cannot overwhelm the damselfish's defenses, because the females will fiercely defend their nests. However, a large group of wrasse are able to do so, and they are rewarded with a gourmet meal of damselfish caviar.[18]

- **Teamwork** – This familiar category of cooperation and synergy can be differentiated from a division of labor, perhaps, in terms of its underlying intent, or purpose. Teamwork refers to the innumerable situations in nature and human societies alike where a variety of labor is combined toward some common goal – when a set of individuals play distinct parts in a larger, emergent whole. The members of a choir, a football team, an orchestra, or the actors in a stage play combine their distinct "roles" to produce a joint outcome.[19]

Teamwork is especially important in the plethora of symbiotic partnerships in the natural world. One of the most extraordinary examples involves a large, single-celled organism called *Mixotricha paradoxa* – so named because its discoverer found that it is indeed paradoxical. Each *Mixotricha* cell involves a collaborative association among at least five different kinds of microorganisms. In addition to the large host cell, there are three external, surface partners, including large spirochetes, small spirochetes, and bacteria. The function of the large spirochetes, if any, is not clear. However, the small hair-like spirochetes, which typically number 250,000 or more *per cell,* cover the entire surface and provide an effective propulsion system through their closely coordinated undulations. Each of these spirochetes is, in turn, anchored by a third symbiotic partner, a rod-shaped bacterium. Finally, inside each host cell there are thousands of spherical bacteria that serve as energy factories, much like the mitochondria in other complex cells. What makes this five-way example of teamwork even more remarkable is the fact that each *Mixotricha* protist is itself involved in a symbiotic relationship. It populates the intestine of an Australian termite called *Mastotermes darwiniensis*, where it performs the vital service of breaking down the cellulose ingested by its accommodating host.[20]

- **Functional complementarities** – This could be viewed as a variation on teamwork. Many synergies involve elements or parts with very different characteristics that complement one another and produce otherwise unattainable collective properties. Lichens provide one example, but one of my favorite examples involves nutrition. One cup of beans, eaten alone, provides the protein equivalent of two ounces of steak. Three cups of whole grain corn flour, eaten by itself, provides the equivalent of five ounces of steak. The sum of the two is seven ounces. But when the two vegetables are consumed together, they provide the equivalent of 9.33 ounces of steak, or about 33% more useable protein. The reason is that their constituent amino acids are complementary. Grains are low in lysine, while legumes are low in methionine. When the two food items are combined, they compensate for each other's deficiencies. In other words, the whole taco is greater, nutritionally, than the sum of its parts.

Functional complementarities can also play an important role in social behavior. One classic example involves the symbiotic partnership between humans and a small bird, appropriately called the honey guide. Humans have been collecting honey for at least 20,000 years, and this highly nutritious natural product has always been prized by hunter-gatherers like the nomadic Boran people, who live in the dry bush country of northern Kenya. However, the search for honey is a major challenge, because honey combs are often widely dispersed and hard to find. They can be hidden in the branches of large trees, or in rock crevasses, or perhaps in an abandoned termite mound. The Boran honey hunters, when they search on their own, often find no hives at all. Even on a day when they do get lucky, they must spend an average of 8.9 hours searching for a beehive, according to the findings of biologists Hussein Isack and Hans-Ulrich Reyer in their classic three-year field study.[21] But if the honey hunters are able to team up with honey guides, the average search time drops to 3.2 hours, or a savings in time and energy of about 64 percent. Once considered to be a myth, or perhaps the unintended result of convergent behaviors, we now know that this legendary example of human-honey guide cooperation is an orchestrated affair. What's in it for the honey guides is that the birds would almost certainly not be able to gain access to the beehives without the help of the honey hunters (only four percent of the time, in the Isack and Reyer study). The honey hunters typically use smoky fires to drive away the bees and then deploy tools to break open the hives. There is also a complementarity in how the hunters and the honey guides divide up the spoils. The hunters take the honey while the birds prefer the beeswax and bee larvae. Nothing goes to waste.

- **Threshold effects** – The colloquial expression "the straw that broke the camel's back" (mentioned earlier) and, more recently, Malcolm Gladwell's "tipping point" involve what are referred to as threshold effects – the many situations where an incremental change results in a radically different outcome (recall *Who Sank the Boat*?).

One everyday example is the way water crystallizes and changes to ice at 32 degrees Fahrenheit (0 degree Centigrade). A classic biological example involves a set of experiments (cited earlier) by Warder C. Allee, a prominent biologist of the 1930s and 1940s. Allee demonstrated that a large number of flatworms (planaria) working together could collectively detoxify a solution that would be lethal to any one of them alone. When a single planarian was placed in a silver colloid solution, its head began to degenerate within ten hours. But when groups of 10 or more planaria were put into the same solution at the same time, they were able to survive without ill effects. Each planarian absorbed a portion of the silver colloid, and their collective efforts reduced the ambient concentration below the toxic level.[22]

- **Augmentation or facilitation** – Another class of synergistic effects involves joint actions that improve the effectiveness of a process. One obvious biological example, mentioned earlier, is catalysts – the enzymes that facilitate biochemical reactions by reducing the activation energy required while themselves remaining unchanged. For instance, the conversion of carbon dioxide and water to carbonic acid occurs some ten million times faster when it's catalyzed by carbonic anhydrase.

  Another important example, also mentioned earlier, involves hemoglobin, one of the most vital molecules in our bodies. Hemoglobin is a complex protein whose four distinct strands (or monomers) are wrapped around a ring-shaped "heme group" including an iron atom. It seems there is a functional interaction between the scaffolding provided by the monomers, the so-called "globin chains," and its internal heme group. Together the parts are able to reversibly bind and then transport molecules of oxygen, carbon dioxide and (we now know) nitric oxide, all of which are involved in the workings of complex biological systems. And that's not all. It happens that the four globin chains are linked together and have collective facilitation properties. When any one chain binds, say an oxygen molecule, it increases the binding affinity of the other three chains, so that each hemoglobin molecule is more likely to load up and transport four oxygen molecules together on its trip through the bloodstream.[23]

A very different kind of mutual facilitation was recently discovered by researchers at the Craig Venter Institute and the University of Southern California. It seems certain bacteria respond to environmental deficiencies of various kinds (like an oxygen shortage in aerobic bacteria) by growing tiny hairs called "nanowires" that conduct electrons. These nanowires enable the bacteria to establish an electrochemical signaling network with their neighbors, which in turn allows them to coordinate and optimize their joint efforts.[24]

- **Joint environmental conditioning** – We now know that this is a very common form of cooperation in nature. A legendary example (again mentioned earlier) is the way emperor penguins huddle closely together in large colonies, sometimes numbering in the tens of thousands, to share heat during the bitterly cold Antarctic winter. By so doing, they are able to reduce their energy expenditures by 20 to 50 percent, depending on where they are in the huddle and the wind direction and speed.[25] In a similar fashion, honeybees, either through collaborative heat production with their bodies or fanning activities with their wings, are collectively able to maintain the core temperature of their hives within a narrow range.[26] And Mexican desert spiders are notorious for huddling together in huge colonies during the hot, dry season, enabling them to reduce individual water loss and decrease the risk of dehydration.[27]

Of course, many organisms modify their environments in more aggressive ways. There is, in fact, a relatively new and important area of research in biology called "niche construction theory," which is focused on how evolution has been shaped by the way living organisms frequently manipulate their environments – from beaver dams and underground mole rat burrows to the dramatic changes that earthworms make in the soil under our feet.[28] The key idea is that these changes may in turn re-shape the selective contexts and the course of evolution itself over time through a process of "reciprocal causation." (There will be more on this subject in Chapter 5, and we will examine the premier example – human evolution – in Chapters 7 and 8.) There is also a new research field in ecology called "ecosystem engineering."[29]

- **Animal-tool "symbiosis"** – Closely related to this is a sometimes underrated but hugely important category of synergistic effects that result from the combined action of a user and some physical object. Thus, some birds use rocks to break open egg shells while others deploy thorns to dig for grubs under the bark of trees. Some chimpanzees use "wands" to fish for buried insects while others use stone anvils and hammers to crack open the proverbial tough nuts. California sea otters are famous for floating on their backs and using rocks resting on their bellies as anvils to break open mussels and other hard-shelled prey. Elephants are especially impressive tool-users. They use their trunks to manipulate a variety of objects – grass, sticks, stones, tree branches, etc. – for various purposes, from cleaning and scratching themselves, to wiping cuts, reaching for inaccessible food, making threat displays, throwing objects at other animals (including humans), and much more.

Of course, we humans are in a class by ourselves. Tools, and an ever-growing repertoire of technologies, have played a major role in shaping human evolution to date, as we shall see (in Chapters 7 through 9), and will no doubt continue to do so in the future. Nothing illustrates better the synergies produced by human tools than the history of agricultural technology in the United States. Back in the eighteenth century, when farmers used oxen or work horses, along with crude wooden plows, hand seeding, harvesting with a scythe, and threshing with a hand flail, one farmer with 400 hours of work was able to produce about 100 bushels of wheat on five acres of land. By 1830, after the introduction of iron plows and other improvements, the labor for 100 bushels of wheat hada dropped to about 250–300 hours. As the nineteenth century proceeded, there were a series of major innovations – steel plows, mechanical seeders, mowing machines, field wagons, and more – which reduced the number of hours for 100 bushels of wheat to 75–90. By 1890, gang plows and harrows, mechanical planters, threshing machines and other new technologies decreased the labor required even further, to somewhere between 40 and 50 hours. Further dramatic improvements in productivity occurred in the early twentieth century, with the introduction of motorized tractors, artificial fertilizers, and other innovations. By 1930, one farmer could produce 100 bushels of wheat with 15–20 hours of work. Each farmer could now feed almost 10 people, compared to about five for a native New Guinea horticulturalist. But this was still far from the end of the story. With the introduction of more powerful tractors, improved plows, disks, seed drills, combines, trucks, new seed varieties, and other changes during the past half century, one farmer nowadays can produce 100 bushels of wheat on three acres (two fewer than his ancestors) with about three hours of labor, less than one percent of what was required in the eighteenth century. Each American farmer can now feed close to 100 people. And GPS satellites, drones, robots and other new technologies promise even more improvements to come. [30] It's the ultimate animal-tool symbiosis.

- **Risk and cost sharing** – This is one of the pillars of social life, both in nature and in human societies. It involves the ability to "economize" by sharing the burdens and costs involved in earning a living. There are innumerable examples of this in the natural world – fish schools, migratory bird formations, synchronized breeding, joint nest-building, collective foraging, and more. Many birds and some animals (like prairie dogs and meerkats) also divide up the job (and the risks) of lookout duty; they take turns scanning the environment for potential predators. In human societies, our insurance industry provides another important example. As Winston Churchill put it, insurance brings "the magic of averages to the rescue of millions."[31] In the same way, the synergies derived from cost sharing are ubiquitous in modern societies. Hotels, restaurants, taxicabs, airlines, cruise lines, equipment rental services and much more all depend on customers who tacitly cooperate in the use, and the costs involved, when they temporarily "rent" something.

- **Information sharing and collective intelligence** – Cybernetic control information plays an indispensable role in coordinating the activities of living systems at all levels. Echoing the classic definition of available energy in physics, control information can be defined as "the capacity to control the capacity to do work."[32] In Chapter 6, where the major transitions in evolution is briefly reviewed, we will explore its vital role much further. Here I want to highlight the cooperative aspect of information.

  Information sharing is one of the more important forms of synergy in the natural world. Indeed, all socially organized species absolutely depend on it. Very often it's a service that can be provided to others at no cost to the possessor, or at an incremental additional cost, while the benefits can be multiplied many times over. The benefits can range from more efficient foraging activities to preserving life itself in a dangerous situation. Although there is now a vast and rapidly growing research literature on the subject, some of the underlying principles were first developed by Darwin in his lesser known but now more widely appreciated book, *The Expression of the Emotions in Man and Animals* (1872).[33]

  Information can take many different forms, from the chemical trails laid down by bacteria and social insects to the vocal signals used by various mammals and birds, and the many kinds of visual displays – such as the waggle dance utilized by foraging honey bees. As suggested earlier, animals not only share information with each other, they frequently engage in pooling data and making collective decisions. This too represents a form of synergy. One well-known category is called "quorum sensing," and it has been observed in (among others) bacteria, social insects, and some colonial species.[34] There is now also a growing body of research on the idea of "collective intelligence" or "swarm intelligence" – smarts that are greater than the sum of the individual parts.[35]

A striking example of such collective decision making in nature is the way a honey bee colony goes about choosing a new home. Bee expert Thomas Seeley calls it "honeybee democracy."[36] Several hundred scouts search for a location that is sheltered, secure, protected from predators, and large enough to meet the colony's needs. Various candidate locations may be identified by the searchers, which are then advertised to the colony, using variations of the waggle dance, in order to recruit more assessments from other searchers. Eventually there is a vigorous debate about the available options and a consensus decision is made for the colony that is almost always the best choice. Seeley calls it "swarm smarts."

Biologists David Sloan Wilson and Edward O. Wilson stressed the potential synergistic value of social information in a recent article on "Evolution 'For the Good of the Group.'" Referring to a phenomenon characterized as "the group mind," Wilson and Wilson observed that "the collective benefits of making a wise decision can be great and the within group costs can be low." [37] (We will have more to say about the synergies associated with collective intelligence in relation to the evolution of humankind in Chapter 8.)

- **Convergent historical effects** – last, and perhaps least appreciated, is one of the most important and pervasive forms of synergy in nature and human societies alike. It involves the daily assault of fortuitous, often unexpected convergent effects that can re-shape our lives and even influence the evolutionary process itself. (The late Stephen Jay Gould wrote extensively on this subject under the headings of "contingency" and "serendipity".)[38] Here synergy and history join hands. Many of the synergies that surround us and impact on our daily lives are unplanned, unconnected and highly context-dependent. They involve two or more distinct causal dynamics that intersect. These dynamics might be, in part, a product of the laws and principles of physics and chemistry, but they can also entail many "accidental" influences with important functional consequences. An obvious example is a highway traffic jam – an unwelcome daily experience for millions of commuters around the world. Another example is global warming, which nobody wants but everyone promotes whenever they light a fire, barbeque a steak, drive a car, turn on a furnace, take an airline flight, or buy products manufactured in modern factories and shipped in containers all over the world. (More on this in the final chapter.) When such historical contingencies result in a major disaster, it could be called the "*Titanic* Effect" – from one of the most horrific ocean-going tragedies of all time.[39]

Every one of the various combinations of labor described above (and there are others as well) has been a driver for Synergistic Selection over the course of time.[40] In effect, the economic benefits may provide important survival advantages, both in absolute terms and in relation to various competitors. The ultimate product of this synergistic dynamic is evolution of complex, integrated *social* systems. They represent an emergent new level of biological organization and synergy – a unit of Synergistic Selection with far reaching consequences. Functionally organized social groups have rightly been called "superorganisms."

The Superorganism

The superorganism concept has a venerable pedigree. It was coined by the nineteenth century theorist Herbert Spencer, although the idea itself is much older.[41] In fact, it was Plato, in the *Republic*, who first likened a complex society to an organism. The whole, including its division/combination of labor, produces synergies that collectively benefit the individual members.

Early on in the twentieth century, a number of biologists, including Warder Allee and William Morton Wheeler (a pioneer student of insect societies), also adopted the concept.[42] However, the term went into a total eclipse in the mid-twentieth century, when neo-Darwinism was in its prime-time. Many biologists of that era embraced the reductionist, gene-centered evolutionary paradigm briefly described in Chapter 3 (with more on this subject in Chapter 5). The superorganism concept became an outcast and was widely criticized as an inappropriate, even "mystical" metaphor, or at best an epiphenomenon without any substance. Biologist George Williams, in his widely influential critique of group selection, *Adaptation and Natural Selection* (1966), dismissed the superorganism concept as a figment of a "romantic imagination" and "not an appreciable factor in evolution."[43]

In the latter twentieth century, however, as evolutionary biologists began to come to grips with the problem of explaining the rise of complexity in evolution, the superorganism term gradually regained admittance to the lexicon. It was Edward Wilson who reintroduced the term, somewhat tentatively, in his ground-breaking volume, *Sociobiology: The New Synthesis* (in 1975). Wilson cautiously broke the taboo in a chapter devoted to so-called colonial species that are intermediates between loose aggregations of individuals and complex, fully integrated organisms. Colonialists include the siphonophores (such as the Portuguese man-of-war), various myxobacteria, slime molds, flagellates, flatworms, coelenterates (like jellyfish), and others. Whenever a colonialist displays a division/combination of labor between reproduction and other functions, much like the reproductive organs and somatic cells in our own bodies, Wilson concluded that "the 'society' can be viewed equally well as a superorganism, or even an organism."[44]

A "Shared Fate"

When group selection advocates David Sloan Wilson and Elliott Sober published a journal article on "Reviving the Superorganism" in 1989, they adopted a viewpoint closer to Herbert Spencer's original functional definition.[45] They wrote: "...the hallmark of an organism is functional organization....We define a superorganism as a collection of single creatures that together possess the functional organization implicit in the formal definition of organism."[46] In other words, a superorganism exists whenever there is an interdependency (a "shared fate," in Wilson's term) that affects natural selection.

In their comprehensive book *The Ants* in 1990, and in a subsequent popularization, co-authors Bert Hölldobler and Edward Wilson stressed that colonies of social ants are also "the equivalent of the organism, the unit that must be examined in order to understand the biology of the colonial species."[47] After describing the complex organization of leaf cutter ants, Hölldobler and Wilson concluded that these ant colonies "do precisely the right thing for their own survival. Guided by instinct, the superorganism responds adaptively to the environment."[48]

Indeed, leaf cutter ants are viewed by some theorists as the "ultimate" superorganism. Because this concept will be important for understanding human evolution (see Chapters 7 to 9), let's take a brief look at the leaf cutter example.[49]

The "Ultimate" Superorganism?

Many millions of years ago, long before our earliest hominin ancestors first appeared, some New World social ants in the higher Attine "tribe" developed a sophisticated form of underground agriculture. The leaf cutter ants create huge subterranean "cities" with extensive tunnel systems and many hundreds of chambers where carefully nurtured, symbiotic fungus gardens produce food for a population that can number in the millions. It's a unique system, and it has allowed the leaf cutters to dominate their local environments. (Recent excavations, using a new method for pouring a concrete mold, have shown that a mature nest might occupy the area of a small house and go 10–20 feet underground.)

A mature ant colony, perhaps five- or six-years old, is a truly impressive organization, with a variety of tasks that are shared among a half dozen anatomically specialized "castes" varying in age and size. (The large "soldier caste," which defends the colony against predators, is some 200 times bigger, in dry weight, than the smallest caste, called "minims.") The various worker castes form into closely coordinated, flexible teams that work together, or in parallel, shifting from one job to another as conditions require.

Members of the youngest minim caste mostly stay inside the nest. Many of them are responsible for nursing and feeding the brood. (The queen lays some 20 eggs per minute, or about 10 million eggs per year. Because many of these eggs are defective; the "bad eggs" are fed to the queen and the brood.) Other minim workers are responsible for tending the fungus garden chambers – cleaning them, removing wastes and parasites, and harvesting the clusters of nutritious swellings produced by the fungi, called staphylae, that provide food for the colony. (We now know that the ants also consume sap from the plants they harvest.)

A "Highway" System

Meanwhile, older, larger workers are busy creating and expanding the tunnel system and carving out new fungus garden chambers. Ultimately there may be hundreds of "pods" that branch off from the main tunnels. Many tons of soil must also be carried off over time and dumped into an exterior mound. Still other ants, the "road workers," are occupied outside the nest creating a "highway" system for the colony – a set of pathways that are cleared of grass and underbrush and impregnated with pheromone chemicals that serve as signals to guide the ants when they are working outside the nest. These pathways are, in effect, an extension of the nest itself, a network of trails that can fan out in all directions and reach a hundred yards or more from the nest. (The benefit is that the trail system increases transport efficiency by some 4–10 times.)

The primary purpose of this extensive trail system is to facilitate the major external activity of the colony – harvesting and transporting fresh leaf cuttings and grasses back to the nest to provide food for the fungus garden. (Over the course of a year, the colony might harvest a square mile of cuttings.) This operation is highly choreographed. The initial leaf cutting task is a demanding, energy-intensive activity, so it is assigned only to the largest workers. The leaf cutters have outsized mandibles and strong head muscles that allow them to use their mandibles like saw blades to slice through thick leaves and cut off large pieces. The job also requires precisely coordinated bodily movements.

The freshly cut leaf segments are then sliced into smaller pieces by other workers, who in turn pass them on to "transporters," workers who carry them back toward the nest in a relay chain with other workers.

Small minim workers may also ride "shotgun" on the leaf segments during the transport operation to defend against parasitic flies that might try to deposit their eggs on the cuttings. Once inside the nest, the transporters hand off the cuttings to still other workers, who move them deeper into the underground labyrinth. There the cuttings are broken down into still smaller pieces that are molded into moist pellets, fertilized with droplets of (nutritious) fecal liquid, and then fed to the fungi.

Colony Hygiene

While the leaf cutting operation and the nurturing of fungus gardens are vitally important priorities, there are other critical jobs that must also be attended to. One involves hygiene. Keeping the colony clean and protecting it against unwanted aliens is a relentless challenge. The nest is often contaminated with foreign bacteria, fungi and other parasites and pathogens. To counteract these threats, workers devote a significant share of their time to removing the intruders, along with inoculating the garden substrate with their fecal liquid, which is laced with various antibiotics, to deter pathogens. They also emit chemicals designed to maintain an acidic pH level of 5. This is favorable for the fungus garden while also serving to deter unwanted pathogens and parasites.

In addition, we now know (thanks to Cameron Currie and his co-workers) that leaf cutter ant colonies seem to depend on the services of a symbiotic bacterium that has been battling a pernicious ant colony parasite (by secreting an antibiotic) in an "arms race" that may go back a very long time.[50] Other evidence suggests that the leaf cutters are actually involved in a four-way partnership. It seems there is another symbiont that assists in breaking down the cellulose in plant cuttings, much like the bacteria that perform a similar service for termites like *Mastotermes darwiniensis* (as noted earlier).

Another regular activity for the workers is the prompt removal of various wastes from the colony, because these can rapidly become toxic. Many of the wastes are transported to empty nest chambers and then permanently sealed up, rather like a nuclear waste dump. Likewise, the potentially lethal carbon dioxide gases produced by the ants and fungi inside the nest must be vented to the outside, a problem that can become critical at times when the colony entrances are sealed up by the workers to prevent flooding during a heavy rain storm. (To partially compensate, the fungi may temporarily slow their growth rate to reduce gas exchange during a storm.)

How Leaf Cutters Communicate

How is all this activity (and much more) orchestrated and coordinated? How are collective decisions made for the colony and its castes, and how does each individual ant know what to do? We now appreciate that information and communication processes play an indispensable role in social insect societies.[51] Any complex social organization depends on having appropriate means for making decisions and passing information among its members (a political system), and we have learned that social insects like the leaf cutter ants have a highly evolved and elaborate communication system. The genes tell the ants what to do, but the when, and the how, not to mention the number of workers that should be involved in doing it, is a dynamic process that utilizes an array of chemical, visual, auditory and tactile signals along with other cues, often in combination.

These impressive communication skills have enabled the leaf cutter ants to utilize many of the different categories of cooperation and synergy described above – a division (combination) of labor, synergies of scale, teamwork, joint environmental conditioning, augmentation effects, functional complementarities, risk and cost sharing, tool use, and collective intelligence. Leaf cutter ant colonies represent a nexus of synergies. The survival and reproductive success of a colony as a whole depends on its multifaceted combination of labor – just as sponges depend on the coordinated actions of their six specialized cell types.[52] A leaf cutter colony is truly a superorganism. (A more detailed account of the leaf cutters can be found in an outtake at my website.)

A leaf cutter colony is also a "collective survival enterprise" and a prime example of Synergistic Selection. In the next few chapters we will take a closer look at this theory and will then apply it to the major transitions in evolution and, ultimately, to the other great example of an evolved superorganism – humankind.

---

[1] Smith 1964/1776, Book 1, Ch. 1, p.7. (For more, see the outtake at my website.) Also, https://books.google.com/books?id=rBiqT86BGQEC&printsec=frontcover#v=onepage&q&f=false

[2] Stigler 1951. (For details, see the outtake at my website.)

[3] Durkheim 1997/1893. Quoted in the Wikipedia article on "Division of Labour" http://en.wikipedia.org/wiki/Division_of_labour (last modified 29 October 2014). See http://sites.middlebury.edu/individualandthesociety/files/2010/09/division-of-labor.pdf For a critique of Durkheim, see Corning 1982. (See also the outtake at my website.)

[4] Bonner 2003.

[5] Anderson and Franks 2001. See also Franks 1989.

[6] Simpson 2012.

[7] Ratcliff *et al.* 2012. McShea and Brandon (2010) have elevated this (presumed) tendency into an evolutionary "law". (For more on their theory, see the outtake at my website.)

[8] Hölldobler and Wilson 2009, p. 84.

[9] E.O. Wilson 2013, p. 121.

[10] Plato 1946/380 B.C., Book 2, 369a,b, 370b,c, 372c.

[11] Smith 1964/1776.

[12] The following is synthesized from Bergquist 1978; Ricketts and Calvin 1985; and Curtis and Barnes 1989.

[13] Smith 1964/1776, Book 4, Chap. 2.

[14] A reviewer of this book in manuscript pointed out that sponge cells are actually totipotent, meaning that adjacent cells could revert to an amoeboid state and then migrate to replace the missing cells. Nevertheless, as an example of functional differentiation sponges still provide a valid example.

[15] See Corning 2003.

[16] Voltaire's famous (mis)quote about God being "on the side of the big battalions" is discussed in an outtake at my website.

[17] Bonner 1988. See also Shapiro 1988.

[18] Dugatkin 1999, pp. 14-15. See also the analysis of synergies of scale in microbial cooperation in Cornforth et al. 2012.

[19] It should be noted that biologists Carl Anderson and Nigel Franks (2001), in their article on "Teams in Animal Societies," used the term as a synonym for the division of labor I prefer to distinguish between the two concepts to highlight the distinction between a task divided and roles that are combined to produce an emergent whole.

[20] Margulis and Sagan 2002; also, Mayr 1974. See also the Wikipedia article http://en.wikipedia.org/wiki/Mixotricha_paradoxa (last modified 18 July 2014).

[21] Isack, and Reyer 1989. This classic study was recently reinforced and amplified by a study among the Yao people in Mozambique. See Spottiswoode *et al.* 2016.
[22] Allee 1951/1938.
[23] Maddox 1990; Hardison 1999.
[24] El-Naggar *et al.* 2010.
[25] Le Maho 1977. Of course, heat-sharing has also been practiced by humans for many centuries. It's even mentioned in the Old Testament.
[26] Gould and Gould 1995.
[27] Von Wagner 1954.
[28] Earthworms dramatically alter the soils they inhabit in a multi-generational process. It's believed that they were originally aquatic creatures and that they have made many adaptations in order to survive in a terrestrial habitat, including major modifications to their environments. See Odling-Smee *et al.* 2003, especially pp. 11-12, 374-376; also, Laland *et al.* 1999.
[29] Wright and Jones 2006.
[29] These data are from the U.S. Department of Agriculture. See especially https://www.agclassroom.org/gan/timeline/farm_tech.htm (accessed 7 December 2014).
[31] From "A Four-Year Plan for England," BBC Broadcast March 21, 1943. See http://www.ibiblio.org/pha/policy/1943/1943-03-21a.html (accessed 7 December 2014).
[32] For further exploration of the concept of "control information," see Corning and Kline 1998a,b; Corning 2005, 2007b. The term "cybernetics" refers to the science of communications and control in goal-oriented systems. A key concept is "feedback" – the capacity to monitor and adjust the behavior of the system in order to attain, or maintain, a desired goal-state. A classic example is a household thermostat. A brief discussion of the widespread misuse of the cybernetic model can be found as an outtake at my website. A lucid introduction to the science of cybernetics and its important contributions can also be found in Capra and Luisi 2014.
[33] Darwin 1965/1873.
[34] See Capra and Luisi 2014. See also, http://en.wikipedia.org/wiki/Quorum_sensing (last modified 9 November 2014).
[35] See especially Bonabeau *et al.* 1999; Eberhart *et al.* 2001; Couzin 2007; Woolley *et al.* 2010; Henrich 2016.
[36] Seeley 2010.
[37] D.S. Wilson and E.O. Wilson 2008. See also Wilson and Wilson 2007; D.S. Wilson 2015; also, Franks 1989; Couzin 2007.
[38] See S.J. Gould 2002. See also the discussion in Capra and Luisi 2014.
[39] Among the many sources on this infamous historic event, perhaps the most comprehensive and up-to-date is the volume by Lynch (1992). (For a summary of the many "what ifs," see the outtake at my website.)
[40] For example, a very different form of synergy was identified by biologist Mary Jane West-Eberhard (2003, pp. 591-592). She noted that there can be times when two distinct

selection processes may reinforce each other. The so-called "hover wasps" provide an illustration. Their helicopter-like behaviors are used for both mating displays and for foraging.

[41] Spencer 1897. Spencer's definition was strictly functional. He spoke only of an analogy between structures and functions (see vol. 1, p. 592).

[42] Wheeler 1928. Biologist Warder C. Allee (1931) also embraced the term in his book, *Animal Aggregations.*

[43] Williams 1966, p. 220, p. 8.

[44] E.O. Wilson 1975, p. 383.

[45] D.S. Wilson and Sober 1989. A defense of the superorganism concept in my 1983 book, *The Synergism Hypothesis*, was not widely noticed. See also Corning 2005.

[46] D.S. Wilson and Sober 1989, p. 339. See also Seeley 1989.

[47] Hölldobler and Wilson 1990; 1994, p. 107.

[48] *Ibid.*, p. 122.

[49] The following discussion is derived from the definitive description in Hölldobler and Wilson 2009. We are deeply indebted to them for a lifetime of work and publications on insect societies. See also Hölldobler and Wilson 1990.

[50] Currie *et al.* 1999; Currie 2001.

[51] See the discussion of this subject in Seeley 1995.

[52] In fact, sponges display several different forms of synergy – functional complementarities, a division/combination of labor, synergies of scale, and even structural (gestalt) synergies. For instance, the shape of the (classic) sponge, with its exit opening located at the top, utilizes physics to help pull water through its cavity.

Chapter 5

A Tale of Two Theories

Fifty years before Darwin published *The Origin of Species*, the prominent French naturalist Jean-Baptiste de Lamarck proposed his own full-blown theory of evolution in his masterwork, *Zoological Philosophy*.[1] The story of what happened to this theory, commonly referred to these days (often pejoratively) as "Lamarckism," represents one of the more shameful episodes in the history of science.

Although Lamarck was generally dismissed and often denigrated by twentieth century biologists for supposedly being misguided about the mechanism of biological change and inheritance, he was actually a remarkable scientist who was greatly esteemed in his day. Born in Picardy in 1744, Lamarck studied briefly for the priesthood, served as an army officer during the Seven Years War, then began the study of medicine but soon switched to botany and later zoology. Ultimately, he won a prestigious appointment to the Museum of Natural History in Paris, where he produced several landmark studies of plants and animals, along with a method for classifying invertebrates that later helped Darwin in his own work. Although Lamarck went completely blind in his last years, he persevered in his research with the help of others. He died a poor man in 1829, long before Darwin's great work was published.[2]

Lamarck is best-known today for his theory about the inheritance of acquired traits. Less well-known is the fact that he was perhaps the first of many evolutionary theorists to propose what later came to be called orthogenesis – evolution via some intrinsic or internal influence. As noted earlier, Lamarck was apparently the earliest "modern" theorist to propose the idea of an inherent evolutionary trend toward complexity.

According to Lamarck, a primary causal agency in evolutionary change is a natural tendency toward continuous developmental progress, energized by what he called the power of life ("*le pouvoir de la vie*"). Likened by Lamarck to a watch spring, the power of life was portrayed as a purely material agency. It involved the crude notion of an internal energy (of which Lamarck distinguished two kinds, "caloric" and "electrical") that operates within organisms through the actions of postulated inner fluids, in a sort of hydraulic manner.

A Change of Heart

Although this simple, mechanical formulation now seems quaintly naïve, Lamarck deserves credit for recognizing the important role of energy in the evolutionary process. But he should be given even greater credit for changing his views about evolution in response to new information. As the late Stephen Jay Gould showed in a major reassessment, Lamarck in his last publication (in 1820) abandoned his orthogenetic model and embraced what amounted to a clear precursor of Darwin's own evolutionary paradigm.[3]

Lamarck's commendable change of heart has generally been overlooked by subsequent generations of evolutionary theorists, along with the fact that he courageously championed the core idea of evolution against formidable political opposition long before it was fashionable to do so and paid a heavy price for it professionally.

But Lamarck is renowned, even infamous, for his thesis that the course of evolution has also been shaped by "habits acquired by conditions" – that is, the direct inheritance of traits developed during the lifetime of an individual organism. The idea did not originate with Lamarck. It was "in the air" at that time in European scientific circles. Moreover, Lamarck (like Darwin) could only guess what the precise mechanism of inheritance might be. He lived long before the discovery of genes and the genetic code.

The long necks of giraffes provided a model for him. Lamarck proposed that this distinctive trait arose because ancestral giraffes had progressively stretched their necks over many generations in order to reach the leaves in the tops of acacia trees, and that these changes were then passed on to their offspring.

Darwin was a Lamarckian

It has been a source of unending embarrassment to modern evolutionary biologists that Darwin himself accepted Lamarck's thesis about the inheritance of acquired characters, which Darwin referred to as "the use and disuse of parts." As Ernst Mayr pointed out, there are at least 12 references to this "mechanism" in *The Origin of Species.*[4] Darwin even cited some possible evidence in its favor. And in his final recapitulation, he listed as one of the "laws acting all around us" the "variability from the indirect and direct action of the external conditions of life, and from use and disuse."[5] Elsewhere Darwin concluded that "I am convinced that natural selection has been the main, but not the exclusive means of modification."[6]

The third, and least appreciated aspect of Lamarck's theory was that he recognized the importance of the living organism itself and its behavior as a distinct causal agency in evolutionary change. Lamarck no less than Darwin understood that functional adaptation to the environment is a problem for any organism. However, the environment is not fixed, Lamarck observed, and if circumstances (*circonstances*) change, an animal must somehow accommodate itself or it will not survive. Changes in the environment over the course of time can thus be expected to give rise to new needs (*besoins*) that in turn will stimulate the adoption of new "*habits.*"[7]

Furthermore, Lamarck asserted, changes in habits come first and anatomical changes may follow. He wrote: "It is not the organs...of an animal's body that have given rise to its special habits and faculties; but it is, on the contrary, its habits, mode of life and environment that have over the course of time controlled the shape of its body, the number and state of its organs and, lastly, the faculties which it possesses."[8] One obvious illustration is the acquired "habit" among ancestral giraffes of eating acacia leaves and its selective survival consequences.

Darwin also appreciated the role of behavior in evolutionary change, but his view of its relative importance was more guarded (with the important exception of human evolution). He concluded: "It is difficult to tell, and immaterial for us, whether habits generally change first and structure afterwards; or whether slight modifications of structure lead to changed habits; both probably often change almost simultaneously."[9] As always, Darwin provided an example:

> Can a more striking instance of adaptation be given than that of a woodpecker for climbing trees and for seizing insects in the chinks of the bark? Yet in North America there are woodpeckers which feed largely on fruit, and others with elongated wings chase insects on the wing; and on the plains of La Plata, where not a tree grows, there is a woodpecker... which never climbs a tree![10]

In sum, there was not a great distance, theoretically, between the mature 1820 version of Lamarckism and the Darwinian vision of evolution, although the two theorists stressed different causal agencies. As the modern scholars Snait Gissis and Eva Jablonka note in a major reconsideration of Lamarck's work, "Selection is central to Darwinian thought, whereas the generation of developmental variations is central to Lamarckian thought."[11]

(Of course, both theorists were guilty of occasionally talking nonsense. For instance, Lamarck supposed that the anger displayed by male ungulates sent inner fluids to their heads where these secretions gave rise over time to horns and antlers.[12] Darwin, on the other hand, has generally been forgiven by his followers for his misguided theory of pangenesis, where he postulated that migrating "gemmules" in the bloodstream might provide a means for transmitting information to the germ line.)

A Theoretical Convergence

However, the ultimate theoretical convergence between Lamarck and Darwin is not how the evolution story played out in the twentieth century. Darwin's theory triumphed and Lamarck was relegated to the crackpot crowd. (Indeed, Darwin's version of the theory also triumphed over that of his co-discoverer, Alfred Russel Wallace, but that's another story.) To be sure, there were various post-Darwinian efforts to resurrect the spirit of Lamarckism – such as the Organic Selection Theory,[13] and the so-called "Baldwin Effect,"[14] which (in essence) Darwinized Lamarck's insight about the role of behavior in evolution. The basic idea was that behavioral changes could influence the selective context and the direction of natural selection. There were also numerous efforts to find support for a Darwinized version of a Lamarckian inheritance of acquired characters, most notably geneticist Conrad Hal Waddington's genetic assimilation theory in the 1950s.[15]

But all this was marginalized by the rise of the science of genetics in the early part of the twentieth century. In a classic (if gruesome) experiment, the pioneer geneticist August Weismann cut off the tails of 20 successive generations of mice without, of course, producing a tailless strain.[16] Other supposed evidence against Lamarck's thesis of acquired characters could be found in the practices of farmers and pet breeders, who routinely dock tails, notch ears, castrate males, spay females, and so on without affecting their offspring. There are also such human customs as circumcision, pierced ears and noses, shaved heads, and various forms of deliberate mutilation, none of which is heritable.

Unfortunately, in the process of rejecting the Lamarckian view of inheritance, the baby got thrown out with the bath water; the role of the organism itself (and its behavior) as an important causal influence in evolution was summarily rejected. Weismann's strident claim that random mutations are the underlying source of creativity in evolution became one of the cornerstones of the nascent science of genetics (along with the seminal theoretical work of Ronald Fisher, J.B.S. Haldane, Sewall Wright, Theodosius Dobzhansky, and others) and, ultimately, of a gene-centered evolutionary theory that dominated much of twentieth century biology.

The Modern Synthesis

This mechanistic, gene-focused paradigm greatly appealed to a young biological science that aspired to mimic the law-governed rigor of the then reigning queen of the sciences (physics). For a time, geneticist Hugo De Vries' "mutation theory" even eclipsed Darwinism.[17] But in the late 1930s, a new theoretical synthesis was achieved that later came to be known as The Modern Synthesis (after the subtitle of biologist Julian Huxley's definitive 1942 book).[18] It was recognized that, if mutations – along with genetic "recombination" in sexual reproduction and other molecular phenomena like genetic drift – are responsible for generating biological novelties, they must subsequently be "tested" in the environment by natural selection. This paradigm was summed up in psychologist Donald Campbell's popular slogan "blind variation and selective retention."[19]

The "Central Dogma"

The Modern Synthesis was reinforced in the second half of the twentieth century by the rise of molecular biology. The Nobel biologist Francis Crick, co-discoverer of the double helix and the genetic code, coined the term "central dogma" – an unfortunate label that he later came to regret – to underscore the then widely accepted claim, following Weismann's lead, that there can be only a one-way flow of biological information from the genome to the organism (DNA → RNA → Protein).[20] This is so, most biologists (wrongly) believed, because the germ line is isolated from the rest of the organism and its environment. So, there can be no feedback, or reciprocal causation.

Accordingly, it followed that mutations and related molecular-level changes – subject to the "approval" of natural selection – are the most important sources of novelty in evolution. As the biologist John H. Campbell put it in a review: "Changes in the frequencies of alleles [in a breeding population] by natural selection *are* evolution."[21]

This reductionist, gene-centric paradigm, renamed neo-Darwinism in the latter half of the last century, achieved a global marketing coup when biologist Richard Dawkins published his best-selling popular book *The Selfish Gene* in the 1970s. Dawkins purveyed a caricature of the evolutionary process that has shaped public perceptions down to the present day. (His book remains popular.) As Dawkins put it, "We are survival machines – robot vehicles blindly programmed to preserve the selfish molecules known as genes."[22] The anthropologist Sherwood Washburn called it "genitis" – the genetic disease.

The biological realities, it turns out, are much more complicated. Over the past few decades the fundamental tenets of neo-Darwinism and the Central Dogma have been convincingly challenged. It seems Lamarck (and Darwin) were correct after all in assuming that organisms are active participants in shaping the evolutionary process. There is now a paradigm shift under way from an atomistic, reductionist, gene-oriented, mechanistic (robotic) model to a systems perspective in which "purposeful" actions and informational processes are recognized as fundamental properties of living organisms at all levels.[23]

For instance, the leading microbiologist James Shapiro in his important 2011 book, *Evolution: A View from the 21st Century,* argues that cells must be viewed as complex systems that control their own growth, reproduction and even shape their own evolution over time. He refers to it as a "systems engineering" perspective. Indeed, there is no discreet DNA unit that fits the neo-Darwinian model of a one-way, deterministic gene. Instead, the DNA in a cell represents a two-way, "read-write system" where various "coding sequences" are mobilized, aggregated, manipulated and even modified by other genomic control and regulatory molecules in ways that can influence the course of evolution itself. "We need to develop a new lexicon of terms based on a view of the cell as an active, sentient entity," Shapiro stresses.[24] Echoing the views of a number of other theorists recently, he calls for "a deep rethinking of basic evolutionary concepts."[25]

"Natural Genetic Engineering"

Indeed, Shapiro cites some 32 different examples of what he refers to as "natural genetic engineering," including immune system responses, chromosomal rearrangements, diversity generating retroelements, the actions of mobile genetic elements called transposons, genome restructuring, whole genome duplication, and symbiotic DNA integration. As Shapiro emphasizes, "The capacity of living organisms to alter their own heredity is undeniable. Our current ideas about evolution have to incorporate this basic fact of life."[26]

The senior physiologist Denis Noble, in a recent paper, argues that all the basic assumptions underlying the Modern Synthesis and neo-Darwinism have been proven wrong. Specifically, (1) genetic changes are often very far from random and in many cases are directed by "epigenetic" (developmental) and environmental influences; (2) genetic changes are often not gradual and incremental (Noble cites, among other things, the radical effects of DNA transposons, which have been found in more than two-thirds of the human genome); (3) an accumulation of evidence for a Lamarckian inheritance of epigenetic influences that has now reached the flood stage; and (4) natural selection, rather than being gene focused, is in fact a complex multileveled process with many different levels and categories of causation.[27]

Evolutionary theorist Eva Jablonka and her colleagues identify four distinct "Lamarckian" modes of inheritance: (1) directed adaptive mutations, (2) the inheritance of characters acquired during development and the lifetime of the individual, (3) behavioral inheritance through social learning, and (4) language-based information transmission.[28] It could be called the extended genome.

In a recent review of the mounting evidence for this Lamarckian view, Jablonka concludes: "The existing knowledge of epigenetic systems leaves little doubt that non-genetic information can be transmitted through the germ line to the next generation, and that internal and external conditions influence what is transmitted and for how long."[29] The developmental biologist Mary Jane West-Eberhard goes even further: "Genes are followers, not leaders, in adaptive evolution."[30]

Evolution "On Purpose"

One of those "leaders" is behavior. What Lamarck called "changes in habits" are now more widely recognized as a major causal agency in evolution.[31] The evolutionary biologist Patrick Bateson likes to characterize behavioral innovation as an "adaptability driver" in evolutionary change.[32]

Indeed, over the past half century the evidence for the important role of learning and innovation in living organisms has mushroomed – from "smart bacteria" to problem solving crows, human-tutored apes, and inventive dolphins. (There is now so much research evidence that some excellent earlier work is being overlooked and forgotten.) Examples are almost endless. We now know that primitive *E. coli* bacteria, *Drosophila* flies, ants, bees, flatworms, laboratory mice, pigeons, crows and ravens, various song birds, guppies, cuttlefish, octopuses, dolphins, whales, elephants, gorillas, and chimpanzees, among others, can learn novel responses to novel conditions, via "classical" and "operant" conditioning.[33] One cynic pointed out that behavioral scientists have only confirmed what pet lovers and circus animal trainers have known for centuries.

Our respect for the cognitive skills of various species has also greatly increased. Animals are often capable of sophisticated cost-benefit calculations, sometimes involving several variables, such as the perceived risks, energetic costs, time expenditures, nutrient quality, resource alternatives, relative abundance, and more. Animals are constantly required to make decisions about habitats, foraging, food options, travel routes, nest sites, even mates. Many of these decisions are under tight genetic control, with "pre-programmed" selection criteria. But many more are also, at least in part, the product of experience, trial-and-error learning, observation and even, perhaps, some insight learning.

Many species are also excellent problem solvers. One classic illustration is biologist Bernd Heinrich's laboratory experiments in which naïve ravens housed in cages quickly learned how to use their beaks and claws to pull up "fishing lines" that were hung from their roosts with food rewards tied to the ends.[34]

Decision-Making in Plants

We have also come to realize that even plants make decisions. In the marine alga *Fucus*, for example, biologists Simon Gilroy and Anthony Trewavas documented that at least 17 environmental conditions can be sensed by these organisms, and that the information they collect is then either summed or integrated synergistically as appropriate. Gilroy and Trewavas conclude: "What is required of plant-cell signal-transduction studies...is to account for 'intelligent' decision-making; computation of the right choice among close alternatives."[35]

Tool-use is an especially significant and widespread category of adaptive behavior in the natural world, from insects to insectivores and omnivores (especially our close primate relatives), and it is utilized for a wide variety of purposes (recall the discussion in Chapter 4). As Edward Wilson pointed out many years ago in his comprehensive survey and synthesis, *Sociobiology*, tools provide a means for quantum jumps in behavioral invention, and in the ability of living organisms to manipulate their environments. Tool-use results in otherwise unattainable behavioral synergies.[36] (We will return to this important insight in relation to human evolution in Chapters 7 and 8.)

Social Learning is Commonplace

Also important, theoretically, are the many different forms of *social learning* through collective stimulus enhancement, contagion effects, emulation, and even some teaching. Social learning has been documented in many species of animals, from rats and bats, to lions and elephants, as well as some birds and fish and, of course, domestic dogs.

For instance, red-wing blackbirds, which frequently colonize new habitats, are prone to acquire new food habits – or food aversions – from watching other birds.[37] Pigeons can learn specific food-getting skills from other pigeons.[38] Domestic cats, when denied the ability to observe conspecifics, will learn certain tasks much more slowly or not at all.[39] And, in a laboratory study, naïve ground squirrels that were allowed to observe an experienced squirrel feed on hickory nuts were able to learn the same behavior in half the time it took for untutored animals.[40]

Not surprisingly, the most potent cognitive skills have been found in social mammals, especially the great apes. They display intentional behavior, planning, social coordination, understanding of cause and effect, anticipation, generalization, even deception. Primatologists Richard Byrne and Andrew Whiten, in their two important edited volumes on the subject, refer to it as Machiavellian intelligence.[41] Social learning provides a powerful means – which humankind has greatly enhanced – for accumulating, diffusing and perpetuating novel adaptations without waiting for slower-acting genetic changes to occur. Though it may still be debated whether or not cultural changes have played a role in the evolution of other species like chimpanzees, there can be no doubt that behavioral and cultural evolution played a major part in human evolution, as we shall see.

The "Pacemaker" of Evolution

Ernst Mayr, in a landmark essay back in 1960, characterized behavioral innovations as the "pacemaker" of evolution. He did not, of course, call it Lamarckism, but this is exactly what it was. At the time, Mayr's contrarian views had only a limited influence. Fifty years later, they are now more widely accepted. Here are just a few excerpts from Mayr's classic paper:

> A shift into a new niche or adaptive zone requires, almost without exception, a change in behavior... It is now quite evident that every habit and behavior has some structural basis but that the evolutionary changes that result from adaptive shifts are often initiated by a change in behavior, to be followed secondarily by a change in structure.... It is very often the new habit which sets up the selection pressure that shifts the mean of the curve of structural variation....
>
> The tentative answer to our question "What controls the emergence of evolutionary novelties" can be stated as follows: Changes of evolutionary significance are rarely, except on the cellular level, the direct results of mutation pressure. The emergence of new structures is normally due to the acquisition of a new function by an existing structure. In [these] cases the resulting "new" structure is merely a modification of a preceding structure. The selection

> pressure in favor of the structural modification is greatly increased by a shift to a new ecological niche, by the acquisition of a new habit, or by both.[42]

This dynamic can be illustrated with a laboratory experiment involving the distinctive behavior of the so-called crossbills, or crossbeaks – birds whose mandibles (the two parts of the beak) seem misaligned because they cross over at the tips. There are about 25 species and subspecies of these birds worldwide, including three in the Galápagos Islands, and it has been known since ornithologist David Lack published his famous study, *Darwin's Finches*, in 1947 that the cross-over trait is adaptive. It allows the crossbills to pry open the tough seed cones of the larch, spruce and pine trees on the islands. In fact, each of the three Galápagos crossbill species is strikingly different from the others, with bills that are specialized for different types of seed cones.

In their laboratory experiment, the researchers Craig Benkman and Anna Lindholm started by trimming the beaks of a group of crossbills (a procedure, like cutting your nails, that does not harm the birds).[43] This treatment effectively incapacitated the birds; it confirmed that the cross-over alignment of the mandibles is essential to prying open the seed cones. In other words, the crossbills' distinctive feeding strategy could not have arisen in the first place without a prior structural variation. However, the experiment also showed that, as the bills grew back, a small degree of crossing-over was just enough to allow the birds to open some of the cones, though very inefficiently, and that their performance progressively improved as the beak tips regenerated.

A Synthesis of Darwin and Lamarck

What this experiment indicated is that, just as Darwin suggested, a "slight modification" (step one) may have opened the door to a new "habit" (step two), which natural selection subsequently rewarded with differential reproductive success (step three). The new habit in turn established a new organism-environment relationship in which the causal arrows then flowed the other way as subsequent beak variations were tested in relation to the new feeding strategy (step four). The new behavior favored a more pronounced (more efficient) crossing-over of the beaks. In other words, in this experiment we can observe a synthesis of the core ideas of Lamarck and Darwin. Both made important theoretical contributions.[44]

A Post-Modern Synthesis?

Many theorists these days are calling for a new, post-modern, post-neo-Darwinian synthesis. Some advocate the adoption of a more elaborate "multilevel selection" model.[45] Others speak of an "Extended Evolutionary Synthesis" that would include developmental processes and Lamarckian inheritance mechanisms, among other things.[46] Denis Noble has proposed what he calls an "Intregrative Synthesis" that would include the role of physiology in the causal matrix.[47]

Whatever label may be applied, it is clear that a much more inclusive framework is needed, one that captures the full dynamics, and the interactions, among the many different causal influences at work in the natural world. We also need to view the evolutionary process in terms of multileveled systems – "purposeful" organizations of matter, energy, and information ranging from genomes to ecosystems. And we must also recognize that the level of selection – of differential survival and reproduction – in this hierarchy of system levels is determined in each instance by a configuration, or network of causes. Indeed, the outcome in any given context may be a kind of vector sum of the causal forces that are at work at several different levels at once.[48]

The core objectives of evolutionary biology over the past 150 years have been to explain both the "how" and the "why" of evolution – both the mechanics and the causal dynamics of the evolutionary process. In the heyday of the Modern Synthesis in the twentieth century, this explanatory framework was often truncated to focus on genetic mutations, sexual recombination, and the mathematics of differential selection (changes in gene frequencies) in an interbreeding population, as noted earlier. This mathematical framework, albeit with many refinements, remains the theoretical backbone of biology to this day.

The fundamental problem with this paradigm is that it explains very little. It's a theoretical "black box." As pointed out in Chapter 2, natural selection (properly understood) is not an external causal agency or a mechanism that can simply be plugged into an equation or conjured up as a blanket explanation for any given outcome in the natural world. It's a metaphor – an umbrella term for a wide-open framework that encompasses whatever specific factors may influence biological continuity and change in any given environment. Equally important, it's no longer tenable to view genetic mutations as the primary source of creativity in evolution. There are many different sources of innovation. In the words of Denis Noble and his co-authors: "DNA does not have a privileged place in the chain of causality…"[49]

Extending the Modern Synthesis

The breathtaking advances that have occurred in the biological sciences over the past half century have shown unequivocally that natural selection encompasses a great many important sub-categories of causal influences. Some of these sub-categories were highlighted in a recent edited volume, *Evolution: The Extended Synthesis.*[50] These include (1) the rapidly expanding work in developmental biology (or Evo Devo as it's often called); (2) the important role of "phenotypic plasticity" in nature; (3) the many ways in which living organisms actively re-shape their environments and influence natural selection (niche construction theory); (4) the exciting work in the new field of genomics, which views the genome as an integrated system; (5) the accumulating evidence related to multilevel selection theory, which is focused on the causal dynamics at different levels of biological organization in the natural world; and (6) a growing list of epigenetic (Lamarckian) inheritance mechanisms that have also shaped the course of evolution over time.[51]

I would add that synergistic phenomena represent another distinct sub-category of causal influences in evolution, in accordance with the Synergism Hypothesis, along with "purposeful" behavioral innovations and new "habits" in living organisms, ranging from bacteria to complex human societies. Richard Watson and Eörs Szathmáry also urge us to view evolution as a cumulative process that conforms to formal learning principles and algorithms.[52] Stuart Kauffman in his latest book invokes the influence of emergent phenomena, especially purposeful "agents" in evolution.[53] Others believe that the cultural transmission of learned behaviors should be treated as a distinct sub-category (see Chapter 8). No doubt still more sub-categories have yet to be identified and illuminated.

An Inclusive Synthesis

Going forward, what is needed, it seems, is a more ecumenical paradigm, one that would provide more of a work plan than a finished product. Perhaps it could be characterized as an Inclusive Synthesis. It should be an expansive, open-ended framework for understanding the evolutionary process and for explaining how, precisely, natural selection "does its work" in any given context (what causal factors influence adaptive changes). It would also represent an ongoing work-in-progress rather than a completed theoretical edifice. Nor would it seek to reduce natural selection ultimately to some simple mathematical formula. The immense complexity of the natural environment, and of life on Earth, confounds our urge to simplify things (although, needless to say, modeling can be a powerful tool for understanding specific aspects of the process). No single discipline (or model) can capture such a complex, multifaceted narrative. In the longer run, our theoretical enterprise will require a synthesis and integration of the many different specialized and rapidly growing areas of knowledge.[54]

In the meantime, the historical process through which these multilevel biological systems have evolved over time can be framed in terms of a sequence of major transitions in complexity – from the very origins of life itself to the emerging global society that humankind is now in the process of creating, for better or worse. And, at every level in this biological hierarchy, we can see the influence of synergy and Synergistic Selection. We will take a closer look at these major transitions in the next chapter.

[1] Lamarck 1914/1809.
[2] See Cannon, 1955, 1959. See also Pietro Corsi's (1988) meticulous reconstruction of Lamarck's life and work. (For more on Lamarck, see the outtake at my website.)
[3] Gould 1999a,b. (For a quote from Gould, see the outtake at my website.)
[4] Reported in Noble 2010. (For the details, see the outtake at my website.)
[5] Darwin 1968/1859, p. 459.
[6] Quoted in Noble 2013.
[7] Although Lamarck was often accused, even by Darwin, of proposing that new habits arise as a result of spontaneous "volition" or "desire", he said no such thing. (For more on what Lamarck really said, see the outtake at my website.)
[8] Lamarck 1914/1809, p. 114.
[9] Darwin, 1968/1859, p. 215.
[10] *Ibid.*
[11] Gissis and Jablonka, eds. 2011, pp. xi-xii. See also Corsi 1988.
[12] Lamarck 1914/1809, p. 122.
[13] Although the term Organic Selection was coined by child psychologist James Mark Baldwin, it should properly be called the Lloyd Morgan/Baldwin Effect. Patrick Bateson (2013) reports that both Lloyd Morgan and Baldwin were anticipated by Douglas Spalding (1873). (For more on this, see the outtake at my website.)
[14] Paleontologist George Gaylord Simpson (1953) coined this term. He chose to define the Baldwin Effect much more narrowly than Baldwin and the other Organic Selection theorists. (For more, see the outtake at my website.)
[15] See Waddington 1962. This history is reviewed at length in Corning 1983; see also Corning 2014.
[16] Weismann 1904.
[17] See https://en.wikipedia.org/wiki/Hugo_de_Vries (last modified 20 February 2016).
[18] Huxley 1942.
[19] D.T. Campbell 1974. To be sure, the Modern Synthesis also succeeded in bringing such life science fields as ecology, paleontology, botany, systematics, and cytology under a gradualist, selection-oriented theoretical umbrella, and it inspired biologists to address such important theoretical questions as just how speciation actually occurs in the natural world and how biological diversity is maintained over time? It was also realized that evolution is a process that occurs in interbreeding populations of organisms, not just isolated individuals.
[20] Crick 1970.
[21] J.H. Campbell 1994, p. 86. As West-Eberhard (2003, p. 143) notes: "If pressed to name the mechanism behind the origin of new traits…most biologists would answer genetic mutations."
[22] Dawkins 1989/1976, p. ix. This famous prefatory remark was only one of many provocative statements in the book. (For more, see the outtake at my website.)
[23] See Corning 2005; Noble 2006; Capra and Luisi 2014.

[24] This quote is from an earlier article by Shapiro 2009.
[25] Shapiro 2011, p xvii. See also Jablonka and Lamb 1995, 2014; Jablonka *et al.* 1998; also, Noble 2006, 2013; Pigliucci and Müller 2010; Bateson and Gluckman 2011; Gissis and Jablonka 2011; Jablonka 2013. These changes were summed up in Shapiro 2009. (For details, see the outtake at my website.)
[26] Shapiro 2011, p. 2.
[27] Noble 2013. See also Noble 2012; Jablonka and Raz 2009. For an update to Richard Lewontin's classic analysis of the various "units" of natural selection, see Godfrey-Smith 2015.
[28] Jablonka *et al.* 1998; also, Jablonka and Lamb 2014. Jablonka and Raz (2009) reported that there were then more than 100 documented examples of epigenetic inheritance in 42 different species. It is now apparent that various forms of epigenetic inheritance are ubiquitous.
[29] Jablonka 2013. Jablonka and Lamb (2014).
[30] West-Eberhard 2003, p. 20. She argues (p. 144) that "the most important initiator of evolutionary novelties is environmental induction." (For an evaluation, see the outtake at my website.)
[31] For examples, see Plotkin, ed. 1988; Weber and Depew, eds. 2003. For a detailed review, see Corning 2014.
[32] Bateson 2004, 2005, 2013.
[33] In the index to their book on *Animal Traditions*, Avital and Jablonka (2000) list well over 200 different species. See also Tagkopoulos *et al.* 2008; also, Breed and Moore, eds. 2010.
[34] Heinrich 1995. Heinrich's (1999) book, *The Mind of the Raven,* provides extensive evidence for the mental abilities of these remarkable birds.
[35] Gilroy and Trewavas 2001. See also the broad treatment by Trewavas (2014).
[36] E. O. Wilson 1975, p.172; see also Beck 1980; McGrew 1992. (See the outtake on tool use in chimpanzees at my website.)
[37] Weigl and Hanson 1980.
[38] Palameta and LeFebvre 1985.
[39] John *et al.* 1968.
[40] Cited in Byrne 1995, p.58.
[41] Byrne and Whiten, eds. 1988; Whiten and Byrne, eds. 1997; see also Gibson and Ingold, eds. 1993.
[42] Mayr 1960, pp. 371, 377-378.
[43] The experiment is described in Jonathan Weiner's (1994) wonderful book, *The Beak of the Finch,* which recounts in detail the long-range study by Peter and Rosemary Grant (and their colleagues and their daughter) in the Galápagos Islands. See pp. 182-184. See also Grant and Grant 2014.

[44] Adam Weiss (2015), in a critical paper on "Lamarckian Illusions," disagrees. However, his definition of Lamarckism is very constricted and distorted, focusing on a single purported example.
[45] E.g., D.S. Wilson 1997a,b; Sober and Wilson 1998.
[46] E.g., Pigliucci and Müller 2010; also, Mesoudi *et al.* 2013; Jablonka 2013; Jablonka and Lamb 1995, 2014; Jablonka *et al.* 1998.
[47] Noble 2013. See also Noble *et al.* 2014. (See the outtake at my website.)
[48] For a discussion of group-level selection, see Farine *et al.* 2015. (Also, see the outtake at my website.)
[49] Noble *et al.* 2014.
[50] Pigliucci and Müller 2010. See also Laland *et al.* 2015.
[51] As Pigliucci and Müller (2010) conclude in their edited volume, "[The] shift of emphasis from statistical correlation to mechanistic causation arguably represents the most critical change in evolutionary theory today" (p. 12). They also stress that the concept of "variation" has been greatly expanded beyond genetic mutations and that natural selection is no longer treated as strictly an external agency but includes internal "generative" changes as well. I go one step further. In fact, natural selection is located in the relationships and interactions between an organism and its environments(s) and the consequences for differential survival and reproduction – as discussed in Chapter 2.
[52] Watson and Szathmáry 2016. See also the comments and the authors' responses in *Trends in Ecology and Evolution,* 2016, 31(12): 891-896.
[53] Kauffman 2008. An in-depth treatment of the role of "agency" – purposeful, goal-related behavior – in evolution can be found in Walsh 2015.
[54] In a detailed examination of several current textbook-length treatments of evolutionary theory, philosopher of science Alan Love (2010) found they are in fact quite inclusive and synthetic and have moved beyond a gene-centered approach.

Chapter 6

The Major Transitions in Evolution

When Maynard Smith and Szathmáry's *The Major Transitions in Evolution* was published in 1995, the well-known biologist Egbert Leigh wrote a review for the professional journal *Evolution* that was off the charts, as the saying goes. "This may be the most important book on evolution since Fisher's (1930) *Genetical Theory of Natural Selection*.... Maynard Smith and Szathmáry take an approach that may change the face of evolutionary theory.... I find this book enormously exciting. It is a decisive turning point."[1] In the two decades since this landmark book was published, "major transitions theory" (as it is now called) has become an important theme in evolutionary biology.

As Leigh noted, Maynard Smith and Szathmáry boldly confronted the fundamental challenge – still avoided by many evolutionary biologists at the time – of trying to explain the emergence of complexity in evolution. As Leigh put it in his review, how do "smaller parts, already tested by the struggle for existence, join into larger wholes that become units of selection in their own right."

Leigh himself had wrestled with a narrower and more generic version of this problem in some of his earlier publications. Leigh asked: How can cooperation for "the common good" arise among selfish genes? His response was that, when there are sufficient "collective benefits," the genome might develop mechanisms to exercise control and suppress cheaters and free riders – a process that he referred to metaphorically as "the parliament of the genes."[2] Thus, Leigh tacitly acknowledged the causal role of synergy – the benefits that arise from cooperation.

Maynard Smith and Szathmáry were more explicit. They advanced a theory that was similar to the one I had proposed in more detail in *The Synergism Hypothesis*, in 1983. In their words: "Cooperation will not evolve unless it pays. Two cooperating individuals must do better than they would if each acted on its own...Behavioural examples are easy to think of, but the principle is relevant at all levels."[3]

Thus, the short answer to the "why" question about complexity – why have complex systems evolved over time – is synergy, and Synergistic Selection. The great achievement of the "master architect" (Nowak's term) has been the progressive evolution of emergent, multilevel systems, culminating (it's fair to say) in humankind. And each major breakthrough in this long and complicated history was a prerequisite, a building block, for attaining an even more inclusive level of organization. To repeat, the arc of evolution bends toward synergy.

An Open-Ended Experiment

Far from being law-like and predictable, however, this trend has always involved a creative, open-ended experiment in which economic criteria (broadly defined) have predominated. The rise of complexity in living systems over time has been shaped by the immediate collective advantages – the functional synergies in a given environment, and these synergies are tested anew in each succeeding generation. As the great biologist Theodosius Dobzhansky characterized it, the evolutionary process is a "grand experiment in adaptedness." It's an experiment that has been going on for perhaps 3.8 billion years, and it has been marked by many traumas, dead-ends, and failures along the way.

But if synergy provides the necessary economic incentives for cooperation and complexity, it's obviously not sufficient. The "how" question is equally important. How were the many structural obstacles to cooperation overcome along the way? Maynard Smith and Szathmáry devoted a great deal of attention to this issue in *The Major Transitions*.

The Problem of "Control"

The challenges to creating and sustaining complexity can be viewed as different aspects of the many-faceted problem of exercising cybernetic control. (Cybernetics is the science concerned with the process of communications and control in goal-directed systems, including living organisms and human-designed machines.)[4] For those evolutionary biologists who are still wedded to the selfish gene model of evolution, the main road-block to achieving cooperation and complexity is the danger of conflict and defection by the "players".[5] But from a broader, functional perspective, the basic problem is cybernetic control, and this is fundamentally an engineering challenge.

"Control Information"

Perhaps most critical is the initial problem of establishing and co-ordinating the relationships among the system's parts. Maynard Smith and Szathmáry, both in *The Major Transitions* and in their follow-on book *The Origins of Life*, showed that new forms of information have played a key role in the emergence of complexity at every level, from RNA and the DNA coding sequences in the genome to pheromone chemical signals in social insects, the evolution of language in humankind, and (now) the binary/digital code of the Internet age.[6] Cybernetic "control information" (see Chapter 4) undergirds both the process of invention/discovery and the ongoing cooperative relation-ships that are needed to sustain complexity (as we shall see).

Another set of cybernetic control problems involves the many different threats to the integrity and stability of any biological process, including especially the potential for errors or mistakes, wear and tear, and environmental hazards of varying kinds. Over time, living systems have evolved a great many mechanisms for dealing with these control problems, from regulatory and repair capabilities at the molecular level to various physiological redundancies, sensory/feedback capacities, developmental flexibilities, and, indeed, the opportunities for renewal and change associated with reproduction itself. Innumerable examples can be found in any introductory biology textbook.

Finally, there are control issues related to the very real (though sometimes overstated) problem of reproductive competition in the natural world, as evolutionary theorists ever since Darwin have stressed. There is a vast and ever-increasing body of theory and research on the many aspects of this challenge, with literally hundreds of analytical models. The structural constraints and facilitators of various kinds are addressed in such familiar theoretical categories as kin selection, group selection, and indirect reciprocity, as well as in the extensive work on the role of policing and sanctions in cooperative relationships.[7] In addition, there are many contexts in which all of the participants depend on the benefits (the synergies) in such a way that cheating or free-riding would undermine those benefits and become self-defeating (recall the rowing model and the paradox of dependency in Chapter 4).

In *The Major Transitions,* Maynard Smith and Szathmáry highlighted many of the structural mechanisms that have evolved over time to cope with these control problems. Some of their examples include the vital role of the chromosome (which puts all of the genes into the same rowboat during reproduction); selection for "honest" meiosis and the elimination of segregation distorters (cheaters) from diploid genomes; the transmission across generations of symbiotic organelles (small, organ-like structures) and other machinery inside the cell exclusively via the maternal line to prevent competition; the generally harmonious cooperative relationships between eukaryotic cells and their symbiotic organelles; the purging by honey bee workers of eggs laid by workers other than the queen; and the suppression by leafcutter ants of reproduction among their symbiotic fungi, except when they are colonizing a new nest.

Updating the Major Transitions

Over the past two decades there have been a number of significant additions and refinements to major transitions theory.[8] These are discussed in some detail in two twentieth anniversary review articles, one co-authored by myself and Eörs Szathmáry in the *Journal of Theoretical Biology* and the other by Szathmáry in the *Proceedings of the National Academy of Sciences.*[9]

Not surprisingly, there have also been a number of criticisms over the years.[10] Some critics complain that the term "complexity" can mean many different things (true) and that Maynard Smith and Szathmáry did not define it clearly enough to support their theoretical purpose. This criticism was addressed in the first of the review articles mentioned above, where we describe two different ways of measuring complexity in relation to the major transitions theory. One is structural in nature and the other is functionally oriented (recall Chapter 3).[11]

Another major criticism is that the various transitions identified by Maynard Smith and Szathmáry do not all have the same properties or structural characteristics. How can you develop a theory that applies to both apples and oranges? One set of critics even dismissed the entire major transitions theme, characterizing it as nothing more than an incoherent series of "miscellaneous transitions."[12] This criticism is myopic, in our view. It's rather like ignoring the fact that apples and oranges are both tree fruits. They may differ in some ways, but they also have some major things in common.

The Common Denominator is Synergy

The most important common property in each of the major transitions is that novel combined effects (synergies) established a new level of complexity and an interdependent new "whole" that became a target of differential selection – i.e., Synergistic Selection.[13] It created a new combination of labor that allowed for a distinct new survival strategy and the creation of new ecological niches. Moreover, each new whole in turn became a part of an even more inclusive whole over the course of time. Later transitions absolutely depended on the earlier ones. From the perspective of the Synergism Hypothesis, this process was driven by novel synergies every step of the way. (I will have more to say on this crucial point below.)

Szathmáry, in his 2015 article updating the major transitions theory, consolidated some items and (tentatively) removed an ambiguous case – sexual reproduction – from the original framework. He also added a new category: the evolution of eukaryotes with plastids, which ultimately gave rise to plants. As Szathmáry explained, these revisions now focus the theory more sharply on the emergence of new levels of reproduction and evolution, often characterized as a transition in "individuality."[14]

For the record, I believe that land plants should also be added to the list. It seems that a symbiotic union between primitive green algae and fungi was the key to how the earliest rootless plants were able to invade and "conquer" the land, perhaps 500 million years ago. Needless to say, this was a major turning point in the history of life on Earth. In fact, some 80 percent of all living plant species even today depend on such "mycorrhizal partnerships."[15]

In any case, the major transitions story is also the story about our own species. We literally embody the emergent products of this multilevel, multifaceted biological hierarchy. We stand at a pinnacle. So, let's take a brief tour of this evolutionary narrative before turning to the evolution of humankind in Chapters 7 through 9.

The Origins of Life: "How" and "Why"

What is life? This provocative question was famously posed to the scientific community by the Nobel physicist Erwin Schrödinger in a series of lectures in 1943. Subsequently published as a small book in 1944, it became an instant classic.[16] Schrödinger asserted that life cannot be defined in purely mechanical terms. It involves a thermodynamic process that he called "negative entropy" (a term that others have shortened to "negentropy").

The term was derived by Schrödinger from the counterintuitive concept of entropy in physics. Entropy is a measure of the tendency of available energy to dissipate and lose its ability to do work. As formalized in the Second Law of Thermodynamics, entropy increases as the potency of energy decreases.[17] However, Schrödinger adopted a broader (some would say corrupted) modern version of the term, which has come to mean any form of physical disorder. So, the term "negative entropy" may also be a convoluted way of characterizing physical order. What Schrödinger was saying, in effect, was that life involves an absence of an absence of order. But more important, Schrödinger drew attention to the fact that living systems require energy to do the work needed to create and sustain themselves. Today this is commonly referred to as "metabolism."[18]

Schrödinger further speculated that, at its most basic physical level, the recipe for life should resemble an irregular "aperiodic" crystal, with information embedded in its physical structure. Here Schrödinger was prescient; he was predicting the discovery ten years later that DNA (and the genetic code) plays just such a role as an information "library" in living systems, along with RNA. In other words, the physical basis of life includes cybernetic "control information" – *the capacity to control the capacity to do work in a thermodynamic, energy-consuming process.*[19] (We will return to this key point below.)

Many years later, the Chilean biologists Humberto Maturana and Francisco Varela identified another important property of living systems. They called it "autopoiesis" – or self-making and self-maintenance. Life has a form of self-created autonomy, they observed.[20] Maturana and Varela did not advance a specific theory about how such properties arose and how they work. Nevertheless, their poetic term has been widely adopted – and sometimes abused – by theorists in other disciplines, from systems theory to sociology and philosophy. (Autopoiesis seems to have become a buzzword that can be used to convey the impression of a deep understanding of life while masking how much we still don't know.)[21] The theoretical biologist Stuart Kauffman and others these days prefer the more familiar concept of "agency."[22]

Chemoton Theory

A more systematic approach to the challenge of explaining life, and how it evolved, was pursued by the little known (until recently) Hungarian theoretical biologist, Tibor Gánti, who set out to define and formally model the minimum conditions required for the origin of life. In his

1971 book *The Principles of Life* (which was not translated into English until 2003), Gánti coined the term "chemoton" (a contraction of the words chemical automaton) for his basic model. In a later, more fully developed version of the model, Gánti identified what he believed are the three essential components of life. They are an autocatalytic, self-sustaining network for metabolism, machinery that provides the ability to control growth and self-replication, and a protective envelope that can enclose this machinery and segregate it from the environment.[23] (Needless to say, Gánti's chemoton model also entails synergy – a combination of labor among three functionally distinct elements.)

The envelope or container element in his model, Gánti suggested (following the work of others before him), might have resembled liposomes – fatty bubbles that form whenever clusters of naturally-occurring lipid molecules are immersed in water.[24] Lipid molecules are hydrophobic at one end and hydrophilic at the other end, which enables them to align themselves and combine forces (synergistically) to create an effective non-aqueous enclosure. A similar role is played in cell membranes today by the so-called phospholipids.

On the other hand, the metabolism element has proved to be more problematic. Over the years, there has been much debate about the challenge of creating a self-catalyzing energy production system. Some theorists think that such an orchestrated set of chemical reactions would have required some kind of template to construct it – an external guiding agency of some sort. Others, notably including Stuart Kauffman, believe that, if the internal dynamics were just right, it could have been a self-directed process. Some recent analyses suggest that the probability of this occurring is better than previously thought.[25] The difficulty is that there is nothing closely resembling a chemoton out there today, so we have only models and simulations, along with some suggestive evidence.

But is it "Life"?

In any case, even if Gánti was correct about the three essential building blocks, it still begs the question: Is this sufficient to be called "life"? Maynard Smith and Szathmáry (and others as well) have said, in effect, "not quite." They argue that the threshold to life can only be crossed when an autopoietic chemoton is able to evolve, when there is open-ended variation that can be differentially selected.

I would set the bar one notch higher even than this. Life exists only when it acquires control information – when it gains the cybernetic capacity to respond to inputs from the environment ("feedback") – the ability to adjust its behavior in response to internal or external changes, in order to achieve or sustain a consistent outcome. In other words, life exists when there is a degree of autonomous agency and adaptability.[26] Evolutionary biologists refer to this property as "teleonomy" – evolved internal purposiveness. Yet many theorists seem to overlook this feature in their models, or simply assume teleonomy. Very few acknowledge that it is, in fact, a foundational structural challenge and a distinctive property of living systems. It transcends the laws of physics, and it has played an increasingly important causal role in evolution, as discussed in Chapter 5.[27] In any case, teleonomy is a global property of the whole and not of any specific list of parts – a synergistic effect.

RNA and Cybernetic Control

In other words, life truly emerged only with the onset of what has been called the "RNA World." Ribonucleic acid and its better-known cousin deoxyribonucleic acid (DNA) constitute one of the three basic kinds of biological molecules (the others are lipids and proteins), and we now know that RNA represents the earliest embodiment of the purposiveness (teleonomy) of living systems. RNA is at once a repository of cybernetic control information and an "executor" that can utilize this information to construct and maintain living systems. In fact, it long preceded the emergence of DNA.

RNA usually consists of a single strand made up of one or more nucleotides (which are themselves complex molecules consisting of a varied mix of adenine, cytosine, guanine, uracil, a ribose sugar and phosphate), and it has some extraordinary, synergistic properties. RNA can store information just like DNA, although less efficiently. It can serve as a template for reproduction like DNA, and, in the bargain, it can catalyze other biological processes the way protein enzymes do.[28] In fact, the earliest so-called "protocells" may have been similar to RNA-based viruses, some of which can still be seen today, or perhaps even the simpler "viroids," as they are called.

In the heyday of the Central Dogma, it was thought that RNA was merely a messenger, carrying information from the supposedly all-important DNA molecules to various locations in embryos and living cells. We now know that RNA is far older, much more versatile, and is quite likely to have played a central role in the early stages of evolution, as it still does in complex organisms today.[29] As Szathmáry notes, "there is ample evidence supporting the view that the RNA world in fact existed."[30] Indeed, as we learn more, we may ultimately conclude that it is still an RNA world.

The Mystery of Life's Origins

What remains very uncertain, however, is how RNA arose in the first place (it's a very complex molecule), along with how early in evolution it made its appearance, and what part it may have played as a source of variation and natural selection. (Its instability compared to DNA suggests that it may well have been a prolific generator of organic novelties.) Even more uncertain is where and how life originated and, equally important, where the energy came from that fueled the process (metabolism).

There have been many different theories about these missing pieces. The long-time journalist and science writer, Suzan Mazur, has published a wide-ranging book about this globe-spanning research enterprise and the many competing theories, called *The Origin of Life Circus.* "No one can yet say what the origin of life narrative is despite the fanfare," she concludes.[31]

What could be called the "amino acids first" theory, the idea that proteins evolved before nucleic acids and RNA/DNA, goes back to the famous Miller-Urey "electric spark," or lightning bolt simulation experiment in 1953. Stanley Miller and Harold Urey, then at the University of Chicago, were able to generate various amino acids and sugars in their laboratory from water, methane, ammonia and hydrogen.[32]

However, the Miller-Urey experiment did not "spark" the next experimental steps toward creating proteins – chains of amino acids called peptides. While amino acids are relatively simple molecules, nucleic acids, peptides and proteins – key structural building blocks in living systems – are orders of magnitude more complicated. A protein typically consists of a long, precisely ordered arrangement of various amino acids (there are twenty different kinds altogether) that is elaborately folded into a three-dimensional structure, and it is the precise composition and configuration of the protein that gives each one its distinctive functional properties. So far as we know, only RNA is reliably able to fabricate proteins (as catalytic ribozymes, and in the ribosomes of modern cells).

How Did RNA Evolve?

Thus, a major theoretical focus recently has been on how RNA evolved. One rather ingenious but controversial theory was proposed by the organic chemist Alexander Graham Cairns-Smith in the 1960s. He suggested that masses of clay crystals, which can assume vastly different patterns, could have served as two-dimensional templates for catalyzing organic molecules. Over time, he theorized, these molecules became independent of their clay vehicles and were self-sustaining.[33] Subsequent attempts to test this theory have not been successful.[34]

"Pan-spermia" – the insemination of the Earth with organic compounds, perhaps from Mars or from meteorites, is another theory that has attracted a following. It was first suggested in the early 1970s by the astronomer/mathematician Chandra Wickramsinghe. He proposed that a possible precursor molecule for life, formaldehyde, might have been formed in the clouds of dust and gases found in space and then later carried to Earth.[35] Some critics have objected that this idea just moves the origin problem into outer space. However, a variety of new evidence in recent years has suggested the possibility that various precursor chemicals, and even some organic molecules, might in fact have been fabricated elsewhere in our solar system, and that there may have been some seeding of the primitive Earth by organic raw materials.[36]

Surface Metabolism

Two of the more intriguing current theories involve what has been dubbed the "metabolism first" hypothesis. One theory traces back to the 1980s. Günter Wächtershäuser, a Munich patent lawyer who also has a degree in chemistry, began to ponder the problem of how complex organic molecules could have arisen in the first place, and he developed a detailed case for what he called the "surface metabolism" theory.[37]

Unlike Cairns-Smith, who relied on the geometry of clay crystals, Wächtershäuser proposed that ancient surfaces with high concentrations of iron-sulfite (or perhaps certain other metallic compounds) might have played a direct catalytic role in initiating a primitive metabolic process. Over time, Wächtershäuser theorized, this might have led to the production of increasingly complex organic compounds, and these, in turn, might have led to more complex catalytic processes.

What gave wings to Wächtershäuser's theory was the discovery in the late 1970s that deep sea hydrothermal vents – volcanic fissures in the ocean floor that spew heated water, dissolved minerals and other materials into the water column – are also incredibly rich ecosystems that teem with a variety of unique organisms, including the now-famous giant tube worms. It seems this ocean bottom food chain is supported at its base by an abundance of what are called chemoautotrophic bacteria – single-celled creatures that can tolerate the high heat produced by the so-called "black smokers" and can feed on the hydrogen sulfide that flows out of them. Although hydrogen sulfide is toxic to most organisms, these remarkable thermophilic bacteria can extract the hydrogen and then use it to convert the carbon dioxide in the surrounding water into organic compounds.[38] The bacteria, in turn, obligingly become the food supply for the next level up in the food chain.

Not surprisingly, the discovery of abundant life on the sea floor inspired Wächtershäuser to suggest that ancient hydrothermal vents, which were perhaps even more numerous and varied in the early history of our planet, might have been the incubators for life.[39] His case was strengthened by the fact that metallic surfaces are associated with these fissures.[40] In the years since Wächtershäuser's theory was first proposed, some elements have been tested successfully, but a major problem remains unresolved. How could these ancestral organisms free themselves from their surface catalysts and become autonomous?

Alkaline Vents

An alternative "metabolism first" theory – one that seems to avoid this obstacle – was proposed in the mid-1990s by geochemist Mike Russell.[41] Russell and his colleagues focused on a very different kind of deep sea vent – one that is not volcanic but is driven by a chemical reaction between seawater and the rock fissures in areas where the Earth's mantle is exposed. The hydrogen freed up by this chemical reaction then interacts with carbon dioxide in the water column (with the help of a catalyst and some energy) to produce an abundance of organic reactions.

These so-called "alkaline vents" also have a labyrinth of warm (not hot) porous compartments where organic processes can be catalyzed and sequestered. Russell and others believe that very similar alkaline vents were the "hatcheries of life," generating ever more complex organic molecules that led in time to an early version of what is known today as the Krebs cycle. (The Krebs cycle has the unique ability both to produce free hydrogen and fabricate the universal energy currency in living systems – adenosine triphosphate, or ATP – and to reverse the process and use ATP to create organic materials.)[42] According to this theory, natural selection then "went to work," favoring the most adaptive among the various organic molecules that were produced, possibly even RNA.

The problem of how these organic materials became free-living, autonomous cells was subsequently addressed by Russell and a colleague, the biochemist Bill Martin, with an ingenious proposal.[43] They singled out another of the universal processes in living systems – respiration based on how energy (ATP) can be generated from a gradient of protons moving across a membrane. (The technical term is chemiosmosis.) As it happens, the alkaline fluids produced by the deep-sea vents can interact with the acidic, CO_2 laden waters of the ocean to produce just such a proton gradient.[44] Thus, a free energy source is readily available in the surrounding medium.

Comets and Chemistry

Another promising theory about the origin of life has emerged recently from the work of biochemist John Sutherland's group at the University of Cambridge in England.[45] They have shown that a pair of simple compounds that were very likely to have been abundant during the early history of Earth – thanks to a steady rain of meteorites from outer space – could perhaps have provided the raw materials for life. Sutherland and his co-workers showed that hydrogen cyanide and hydrogen sulfide, together with ultraviolet (UV) radiation, might have been capable of producing all three of the essential organic molecules – nucleic acids, amino acids, and lipids – perhaps with the help of different metallic catalysts. In short, all of the basic ingredients for life, along with an energy source, could have been invented more or less simultaneously in different locations on the Earth's surface and might then have found their way into that "warm little pond" famously envisioned by Darwin, where life could then be synthesized.[46]

Or perhaps it was a "hot little pond." New studies by biochemists Richard Wolfenden and Charles Carter suggest that there was possibly a close linkage between the evolution of RNA and proteins.[47] Wolfenden and Carter argue that there was an interaction – a synergistic co-creation of nucleotides and amino acids/peptides that was facilitated by the fact that the Earth's environment was then still quite hot. "In our view, it was a peptide-RNA world, not an RNA only world," Carter concludes.[48]

All these new discoveries make the origin of life seem more comprehensible. However, the additional "giant leap" (to borrow astronaut Neil Armstrong's punch line from his historic moon landing) from chemistry to teleonomy (and the synergy of an RNA + proteins + lipids world) is still an unsolved mystery. Whatever turns out to have been the actual pathway, it is increasingly likely that life was, after all, invented here and that it was, above all, a synergistic "combination of labor."[49] Life has been synergistic from the get go. In fact, it seems likely that life is not even unique to our planet. However, this momentous origin event was only the first of many further transitions in evolution, for life on Earth is still in the process of inventing itself, as we shall see.[50] We need to hold on to this thought as we take a brief look at the next major step in the evolutionary process.

The Rise of the Prokaryotes

Not so many years ago, the prokaryotes (bacteria and their cousins, called archaea) had a bad name. It was widely assumed that bacteria were the enemy – the bearers of terrible diseases like cholera, syphilis, leprosy, anthrax, pneumonia, tetanus, and others. We now know that the picture is decidedly more mixed. Bacteria (and archaea) can also do a great deal of good, and they deserve some superlatives for their extraordinary accomplishments.[51]

For starters, the origins of the prokaryotes can be traced back at least 3.7 billion years. They were the first complete single celled organisms (so far as we know), and they are still thriving today. They are also by far the most productive form of life on Earth, with an estimated total biomass that outweighs all the rest of the world's fauna and flora combined. In every gram of topsoil there are some 40 million bacteria, and our bodies support an estimated 40 trillion of them, many more than our estimated 30 trillion body cells. (Fortunately, they weigh altogether only a few pounds, and they are mostly mutualistic symbionts or free-riders.[52])

Prokaryotes are also highly inventive and adaptable. Some can survive even in the most extreme environments – ice packs, boiling water, nuclear wastes, garbage dumps, sewers, deep inside the Earth's crust, and in manned spacecraft in outer space, not to mention all those thermophiles in the ocean depths. Among their many skills, bacteria can produce antibiotics, treat sewage, recycle organic wastes, help clean up oil spills, assist in mining minerals, and produce illumination for various animal partners. They also invented photosynthesis, nitrogen fixing, glycolysis, fermentation, cellular damage repair, and a gene-sharing conjugal process analogous to sex. They can even synthesize at least 27 different kinds of minerals – notably including magnetite, which they can use as a compass for navigation.[53]

Profligate Mutualists

Bacteria are also profligate mutualists and play a vital role in sustaining many other species. Virtually all ruminant animals, including some 2,000 termites, 10,000 wood-boring beetles, and 200 species of artiodactyla (deer, camels, antelope, etc.) depend on symbiotic bacteria (and/or fungi and protoctists) to break down the tough cellulose fibers in plants and convert them into a useable form (as noted earlier).[54] Many plants rely on *Rhizobia* – bacteria living in the soil – to fix nitrogen compounds for their roots. A great many of the 25,000 or so different lichen partnerships depend on cyanobacteria for photosynthesis and energy production. Even humans benefit from the 1,000 or more different varieties of bacteria that typically inhabit our digestive systems. Our helpful symbionts can synthesize Vitamin K, folic acid, and biotin, as well as convert sugars, ferment carbohydrates, and help to defend the gut against invading parasites and pathogens.[55]

By far the most important accomplishment of bacteria (and archaea), though, is that they invented the basic machinery of life as we know it. A typical bacterium, which of course cannot be seen by the naked eye, is contained within an outer wall and membrane, and (we now know) a set of internal structural supports called a "cytoskeleton." The interior, or "cytoplasm," typically has no nucleus or organelles, yet there is a clear division/combination of labor among various compartments, including a circular chromosome housing the cell's genome and its DNA, as well as ribosomes where proteins are manufactured.

Bacteria also utilize various strategies for the all-important task of acquiring energy, ranging from aerobic and anaerobic (oxygen-free) respiration to fermentation, photosynthesis, and the capture of free floating protons as a fuel source. Many bacteria also have external flagella that provide mobility for the cell, and they have various ways of sensing and responding to their environments. They typically communicate with one another using various chemical signals and with physical contact.

The prokaryotes also pioneered in exploiting the synergies that can be achieved through social cooperation. The vast majority of bacteria live together in colonies that biologists refer to as biofilms, or microbial mats, where there may even be a social division/combination of labor for everything from food production to reproduction.[56]

"Living Rocks"

One notable example is the so-called "stromatolites" – rocky domes of various sizes that litter seashores from the Bahamas Islands to Shark Bay, Australia. Often referred to as "living rocks," many of these stromatolites are composed of layers of bacteria (and algae), along with inorganic materials (sand, gravel, mud and other particles) that have been deposited by wave action over many years and cemented together by bacterial activity. Some of these remarkable structures are still living, but many others are ancient fossils. The earliest outcroppings date back about 3.4 billion years.[57]

What makes the stromatolite colonies so distinctive is that they use collective action to create a synergistic system – a combination of labor – that can resist the destructive effects of wind, water, waves and ultraviolet radiation. The top layer of the stromatolite typically consists of dead cells that serve the community by shielding the next layer – the energy-producing photosynthesizers – from the damaging effects of radiation. Below that are layers of producers and consumers, along with various channels to allow for the circulation of nutrients, enzymes and wastes. And below that are limestone and the hardened corpses of many previous generations of bacteria. The stromatolite community also has a sophisticated chemistry, utilizing both anaerobic and aerobic processes. And when the sediment at the crown becomes too thick, the next layer below migrates upward, adding a new level to the structure.

Pack Hunters

Bacteria also utilize a variety of different cooperative (synergistic) strategies for acquiring nutrients, including even hunting in "packs" just like our own ancestors. Thus, large colonies of the predatory bacteria *Myxococcus xanthus* (mentioned earlier) forage together and can jointly engulf and consume much larger prey than any one or a few of them could do alone. Equally important, they collectively secrete digestive enzymes in threshold concentrations that would otherwise be dissipated in the surrounding medium. It's a classic example of a synergy of scale. *Myxococcus* colonies have even developed a crude reproductive division/combination of labor. They produce specialized reproductive pods that hang from the tip of each "branch" of their colorful, tree-like colonies.[58]

But perhaps the most significant take-home lesson regarding bacteria is that their colonies defy the selfish gene, nature-red-in-tooth-and-claw model of evolution. It turns out that they are promiscuous gene sharers (the technical term is "horizontal gene transmission" or HGT, as noted earlier), along with engaging in a form of sexual reproduction. This gives bacterial colonies two big advantages. One is genetic redundancy when it comes to coping with translation errors and damage repair, and the other is the ability to adapt very quickly to new challenges and threats. Bacteria can typically reproduce themselves in about 20 minutes, and a single bacterium is capable – at least in theory – of producing billions of copies of itself in the course of a day. In fact, it has been estimated that gene-sharing colonies are able to cope with environmental threats like human antibacterial agents some 500 times faster than any individual bacterium could do.[59]

In sum, bacteria have exploited the potential for synergy in many different ways, and these synergies have played a major part in their enduring success. A bacterial colony is a product of Synergistic Selection. Bacterial systems survive and reproduce – and evolve – as interdependent "wholes." They invented the primordial "collective survival enterprise."

The Emergence of Eukaryotes

The emergence of the eukaryotes – complex single-celled organisms with their genes contained in a sequestered nucleus along with an array of specialized internal organelles – is unquestionably one of the major transitions in evolution, regardless of how you define the term.[60] What sets the eukaryotes apart is that they have been able to fully exploit the synergistic potential of a division/combination of labor at the cellular level and achieve new collective capabilities that bacteria, for all their many talents, clearly lack.

The typical eukaryote is a microscopic miracle – a highly complex network of activities within and between more than a dozen specialized organelles that share in the task of enabling the cell to grow, adapt, repair and reproduce itself. The cell membrane, the internal cytoskeleton, the chromosomes, nucleolus, ribosomes, Golgi apparatus, endoplasmic reticulum, and the other internal structures perform various specialized roles and interact with one another in very precise ways. Even the nucleus turns out to be an immensely complex system (a "command center"). But the key to it all, perhaps, is the role played by the mitochondria, ancient symbionts that provide most of the eukaryotes' energy.[61] (In plant cells, the chloroplasts that specialize in photosynthesis are also of prime importance; they capture most of the energy that the cells use, and they too are derived from ancient symbionts.)

Prokaryotes don't have mitochondria. Many rely for their energy needs on a process called glycolysis, which produces molecules of energy-rich ATP as their basic energy currency. Glycolysis is an exceedingly complicated process, utilizing glucose molecules (a six-carbon sugar) as a raw material. In fact, glycolysis involves no less than ten distinct stages, each catalyzed by a different enzyme. It's yet another example of a complex (synergistic) combination of labor. However, the energy derived from glycolysis is relatively skimpy. Just two molecules of ATP are produced from each glucose molecule, which may then yield about 143 kilocalories of energy per mole. The rest goes to waste in the form of pyruvic acid. Because there is this energy constraint, bacteria are greatly restricted in what each one is able to do.

A Synergy of Scale

Mitochondria, happily, feed on pyruvic acid (and oxygen). Using an even more complex two-stage process involving the Krebs Cycle and electron transport, mitochondria are able to extract as many as 30–34 more molecules of ATP from the pyruvic acid residues, boosting the energy-conversion efficiency of the combined metabolic process from less than 5% to more than 40% and the energy yield to as much as 686 kilocalories per mole.[62]

Equally important, the synergy provided by the mitochondria can be multiplied many times over – a hugely significant synergy of scale. Each eukaryotic cell may have hundreds, or even thousands of these energy-producing organelles, depending upon its particular functional needs. Liver cells, for example, have about 2,500 mitochondria, and muscle cells may have many more times that number. So, the synergistic combination of glycolysis and respiration by the mitochondria, coupled with a synergy of scale, gives the eukaryotes capabilities for growth and other "technological" improvements that would otherwise be impossible. (Eukaryotes are, on average, some 10–15,000 times larger than a typical bacterium.) It's like going from a horse and buggy to a 400-horsepower driving machine.

But perhaps most significant is the fact that the eukaryotes' vastly larger size and more sophisticated technology allows them to become specialists in an even larger, multicellular combination of labor. The obvious analogy in human societies is the development of energy technologies based on fossil fuels (coal and oil), which powered the industrial revolution and all the other technological achievements in human societies ever since (see Chapter 9).

The Paradox of Dependency

On the other hand, the breakthrough in energy production achieved by the eukaryotes, perhaps 1.5–2 billion years ago, had a collateral consequence – a trade-off that we have met before. The price paid for this new level of synergy in living systems was a greater degree of interdependency. The symbiotic parts became dependent upon one another, and their fate became tied to the overall performance of the whole.

The once-independent mitochondria provide a case in point. They still reproduce themselves separately from their host cells, utilizing their own DNA. However, they can't survive independently. They depend entirely upon the raw materials (pyruvic acid, oxygen and other nutrients), and the protection supplied by their hosts. In fact, most of their DNA has been transferred to the cell nucleus. A similar interdependency is found in many other "obligate" symbiotic partnerships in nature and, as we shall see, in the human species as well.

One other major eukaryote invention, less well known but also important, should be mentioned. Prokaryotic bacteria obtain external nutrients by secreting digestive enzymes to break down passing food particles in the surrounding environment. The products of this diffuse digestive process must then be absorbed one molecule at a time through the cell wall. As Maynard Smith and Szathmáry pointed out, this is a very inefficient and wasteful method. Eukaryotes, on the other hand, are able to engage in "phagocytosis." They form pockets in their cell membrane that can engulf and "swallow" passing food particles. These pockets (vacuoles) then fuse with internal organelles (lysosomes) that contain digestive enzymes. Nothing is wasted. Our own digestive systems are descended from this breakthrough biotechnology.[63]

Symbiogenesis

The origin of eukaryotes is a subject of intense debate these days, but, whichever scenario proves to have been the actual route, it involved a cooperative survival strategy and Synergistic Selection. Equally important, it was a prime example of a Lamarckian process (properly understood) in which behavioral innovations played a key role, as discussed in Chapter 5.

It was a Russian botanist Konstantin Mereschkovsky who originally proposed, back in 1905, that the chloroplasts were descended from free-living bacteria and that a beneficial symbiosis explained their appearance in eukaryotes.[64] As noted earlier, Mereschkovsky coined the term "symbiogenesis" to characterize this evolutionary development.

Mereschkovsky's hypothesis – and related work on symbiosis by other Russian scholars – was initially presented as an alternative to Darwin's theory, so it was rejected and remained largely unknown in the West at that time. But in the 1920s and 1930s, another Russian theorist,

B. M. Kozo-Polyansky, recognized that symbiogenesis could also be compatible with Darwinism: "The theory of symbiogenesis is a theory of selection relying on the phenomenon of symbiosis," he wrote.[65] Unfortunately, Kozo-Polyansky's works were published only in Russian and appeared at the height of the Stalinist era, so he was virtually unknown in the West until the 1990s.

Symbiogenesis was also proposed independently in the 1920s by an American biologist, Ivan Wallin, who published the idea in a volume called *Symbionticism and the Origin of Species.* Wallin's "rather startling proposal," as he disarmingly characterized it, was that "bacteria, the organisms which are popularly associated with disease, may represent the fundamental causative factor in the origin of species."[66] Among other things, he claimed that the mitochondria in eukaryotes could be grown independently of their host cells (a dubious proposition). When his theory was vehemently attacked by his peers, Wallin dropped the subject and it was soon forgotten.

A Revival of Symbiogenesis

Only in the 1970s did the idea of symbiogenesis begin to gain favor, thanks to new findings and the more solid case developed by biologist Lynn Margulis. At first Margulis was also ridiculed for defending the idea, but she persisted as the evidence in favor of it mounted (especially important was the then new technology for analyzing DNA "fingerprints").[67]

Today, it is generally accepted that the mitochondria and the chloroplasts did in fact arise through symbiogenesis, and the Nobel biochemist Christian de Duve, biologist Thomas Cavalier-Smith and others think the peroxisomes did too.[68] (Peroxisomes can buffer the cell against oxygen contamination, so it is believed that they may have been the first to join the team and played a crucial role in paving the way for later alliances.)

One remaining dispute concerns Margulis's proposal (again foreshadowed by a forgotten Russian predecessor) that the centrioles and kinetosomes – specialized structures with a distinctive 9(3)+0 architecture that play a vital role in cell motility – also originated via symbiosis, perhaps among the ancestors of modern spirochetes.[69] Another debate concerns Margulis's thesis that the nucleus itself may be a product of symbiogenesis, although the accumulating evidence seems to weigh against this idea.[70]

There is no definitive explanation for how, and why, such symbiotic/synergistic mergers came about in the first place. Several alternative scenarios have been suggested. Many theorists favor a predatory, "infection" model. In the case of the mitochondria, for example, ancestral purple bacteria are presumed to have been invaders that came to prey on their future protist hosts and only later discovered the benefits of mutualism. Other theorists, including Cavaliar-Smith and de Duve, think the key development was phagocytosis – the loss of a rigid outer cell wall and the ability to swallow and digest food internally. In this scenario, the eukaryotes' future symbionts were first ingested as prey, not predators.[71]

A "Fateful Encounter"

A third school envisions that there was a "fateful encounter" between bacteria/archaea that was symbiotic from the start. Lynn Margulis was a long-time champion of this model. One intriguing scenario, proposed by Bill Martin and Miklós Müller, is called the "Hydrogen Hypothesis," and it is supported by a variety of genetic and biochemical data.[72] Martin and Müller believe a mutually-beneficial association developed between an ancient hydrogen-producing bacterium and a "methanogen" – an anaerobic microbe that can utilize hydrogen to extract energy and make sugars, leaving methane as a waste product.

The idea came to Martin one day when he was viewing a modern analogue, a one-celled eukaryote called *Plagiopyla*. These protists have internal hydrogen-producing organelles called hydrogenosomes, which are surrounded by hydrogen-consuming symbiotic bacteria. Martin and Müller believe that these organelles and eurkaryotic mitochondria might both trace their ancestry to a similar partnership. Additional support for their hypothesis comes from the recent research showing that mitochondria are a diverse family, with four distinct types that have both aerobic and anaerobic variants.[73]

A possible resolution to the debate about how the eukaryotes arose may have been discovered by biochemist Nick Lane. He points out that phagocytosis is energetically a hugely expensive capability, and it is very unlikely that this trait (or any of the other expensive traits that characterize eukaryotes) could have evolved in energetically-constrained bacteria. It was the abundant new energy resources provided by the mitochondria that opened the door to the evolution of other new capabilities, and to greater complexity in living systems.[74]

Whichever may in fact have been the pathway that led to eukaryotes, it is clear that synergy, and Synergistic Selection undergirded the process. The benefits for each of the participants vastly outweighed the costs, and the cooperative relationship was reinforced by effective deterrents to cheating (recall the "corporate goods" model in Chapter 3). This synergistic new survival strategy in turn laid the foundation for the next major transition in evolution – the emergence of multicellular organisms.

The Rise of Multicellularity

Multicellular organisms represent, in effect, a cooperative (synergistic) combination of labor. In fact, multicellularity may have arisen independently at least nine times during the course of evolution, although only a few of these initiatives persisted and led to even greater complexity. As biologist David Queller has pointed out, these transitions involved two distinct pathways, which he characterizes as "fraternal" and "egalitarian."[75]

In the former case, all of the participants are closely related, and this facilitates the emergence of a reproductive division of labor (separation of the germ line), as well as the development of functionally differentiated "parts." In the egalitarian case, the pathway is more "democratic" in the sense that each of the participants is genetically unrelated and retains the freedom to reproduce independently, while providing complementary functions for a new partnership. One obvious example of the latter pathway is the mitochondria that reproduce separately in eukaryotic cells. Another example, as we shall see, is found in humankind. However, it appears that the rise of multicellular organisms followed the fraternal route between close relatives.

The economic incentives for multicellular cooperation include many opportunities for achieving synergies. Consider, for example, the volvocines, a primitive order of aquatic green algae that form tight-knit colonies resembling integrated organisms. The volvocines have been popular with students of evolution ever since the nineteenth century, because they seem to mirror some of the intermediate steps toward more complex multicellular organization. The smallest of these species (*Gonium*) have only a handful of cells arranged in a disk, while the *Volvox* that give the volvocine line its name may have some 60,000 cells in the shape of a hollow sphere that is visible to the naked eye. Each *Volvox* cell is independent, yet the colony-members collaborate closely. For instance, the entire colony is propelled by a thick outer coat of flagella that coordinate their exertions to keep the sphere moving and slowly spinning in the water – a synergy of scale.

Some of the other kinds of synergy in the *Volvox* were documented in a study many years ago by the biologist Graham Bell, and in more recent studies by Richard Michod.[76] The largest of the *Volvox* colonies have a division of labor between a multicellular body and segregated reproductive cells. Bell's analyses suggested some of the benefits. A division of labor and specialization facilitates growth, resulting in a much larger overall size. It also results in more efficient reproductive machinery (namely, a larger number of smaller germ cells). The large hollow enclosure in *Volvox* also allows a colony to provide a protective envelope for its daughter colonies; the offspring disperse only when the parental colony finally bursts apart.

Another Synergy of Scale

There is one other vitally important synergistic benefit in *Volvox* – a synergy of scale. As documented by biologist John Tyler Bonner in his book *Why Size Matters: From Bacteria to Blue Whales*, larger size can have many advantages.[77] Indeed, Bonner holds that larger size generally evolves first, because there are immediate synergies, and that other collective benefits may follow. In the case of *Volvox,* for instance, their larger overall size results in a much greater survival rate than in the smaller *Gonium*. It happens that these planktonic algae are subject to predation from filter feeders like the ubiquitous copepods, but there is an upper limit to the prey size that their predators can consume. So, the larger, integrated, multicellular *Volvox* colonies are virtually immune to predation from filter feeders.

Another striking example of synergy among colonialists can be found in the Portuguese Man-of-War, a marine invertebrate that looks a bit like a jelly fish when floating on the surface but is in fact a "siphonophore" – a filter feeder.[78] Legendary for their stinging tentacles, each Man-of-War colony represents a collection of individuals that are attached to one another and are physiologically integrated to the extent that they are not capable of independent survival. Their primitive partnership includes a gas-filled bladder that keeps the colony afloat while three different kinds of specialized polyps – long tendrils that may extend 30 feet or more underneath the bladder – divide up several other tasks. One set of polyps is responsible for stinging and capturing passing prey (small fish and other small creatures). A second set of polyps digests the catch and distributes the products, while a third set forms an organ that is responsible for reproduction. It's a very simple but a highly effective combination of labor. Although the Portuguese Man-of-War lacks a propulsion system, it is found in warm waters around the world.

Multicellular Organisms

Synergies of a similar kind can also be seen in functionally integrated, multicellular organisms, from sponges to leaf cutter ants and blue whales. Take *Homo sapiens* as an example. The human body represents an extraordinary combination of labor, with an estimated 30 trillion cells consisting of about 210 different specialized cell types that are organized into an enormously complex system of differentiated parts. The system is held together by an intricately articulated, flexible skeletal structure and a biologically sophisticated outer envelope, as well as about 650 skeletal muscles (depending on how you count) and literally billions of tiny smooth muscles, along with ligaments, tendons, and a dense network of channels for internal communications and transport services.[79]

All of this elaborate physical structure in humankind is sustained by 10 specialized organ systems with some 50 major parts and many thousands of smaller parts.[80] Add to this our remarkable brain, itself an enormously complex system consisting of some 180 different functional areas (according to some new research), with roughly 85 billion densely interconnected neurons.[81] The many parts of this exceedingly complex system and its individual cells work together seamlessly (for the most part) in a vast combination of labor. And all the parts of this remarkable system are interdependent; they must all survive together in the same "rowboat." As Szathmáry expresses it, "Ultimately, what allows for organism formation from lower level units is a high level of cooperation and a low level of realized conflicts."[82] In short, every move we make, and every thought that comes to mind is an exquisitely orchestrated synergistic effect.

Transitions versus Inventions in Evolution

Before turning to the role of synergy in the evolution of humankind, there is one concern about the major transitions paradigm that should be mentioned. The original definition of a major transition by Maynard Smith and Szathmáry was dualistic; it assigned equivalent importance to functional developments and to key innovations in the storage and utilization of information. I prefer to view such informational changes as "major inventions" – technologies that facilitate but do not themselves produce a major transition in the *strict sense* of a synergy-driven change in the level, or unit of selection. Cybernetic control information is a necessary functional element of every transition, minor and major.

To illustrate the problem, Eva Jablonka and Marion Lamb have advanced the thesis that the development of the nervous system should be added to the list of major transitions in evolution. They argue that "The evolution of a nervous system not only changed the way that information was transmitted between cells and profoundly altered the nature of the individuals in which it was present, it also led to a new type of heredity – social and cultural heredity – based on the transmission of behaviorally acquired information."[83]

I would argue that this important improvement in the informational capabilities of living systems was *not* a major transition but rather a major invention. For one thing, learning is not confined to organisms with brains and nervous systems. We now know that even *E. coli* bacteria can learn to anticipate future events.[84] Control information – defined once again as "the capacity to control the acquisition, disposition and utilization of matter/energy in living systems" – arose with the emergence of RNA (as noted earlier). Over time, informational capacities have evolved in lockstep with every significant increase in biological complexity, and it certainly played a role in all the major transitions. But it does not define them.

An "Egalitarian" Model

Thus, I will (slightly) disagree with Maynard Smith and Szathmáry, and others, who have justified including human evolution on the list of major transitions in part because of our unique verbal communications system and the many different ways it has empowered our species.[85] Yes, human evolution was a major transition, but for a different reason.

I will argue (along with some others) that our ancestors created a unique new unit of selection in evolution – large-scale cooperative social relationships and a complex, multileveled combination of labor (interdependent economic systems) composed of *non-kin* who have retained their individual reproductive capacities (in accordance with Queller's "egalitarian" model) rather than surrendering this function to a "queen" or to some specialized reproductive entity.[86] (Theoretically, this could still happen in the future – think harems and eunuchs – but for the present it's a wildly improbable idea.)

Perhaps the currently popular term "ultrasociality" could be used to characterize the distinctive properties of the human superorganism.[87] As David Sloan Wilson expressed it, "We are evolution's latest major transition. Alone among primate species we crossed the threshold from groups *of* organisms to groups *as* organisms."[88]

Equally important, I will propose (along with others) that the evolution of humankind represents a culmination (and a synthesis) that combined the growing influence of behavioral innovations, increasingly potent technological inventions, a social division (combination) of labor, and both epigenetic and cultural inheritance.

Over time, this synergistic package has supercharged and accelerated the evolutionary process itself. Language was a product of this process, as well as a key facilitator, but it was the economics – the synergies derived from social cooperation and our many novel tools and technologies – that drove the evolution of humankind. The path that our species has followed was shaped by Synergistic Selection. I will elaborate on this theme in the next three chapters.

---

[1] Leigh 1995.

[2] Leigh 1977, 1983, 1991.

[3] Maynard Smith and Szathmáry 1999, pp. 22-23.

[4] See Corning 2014. The science of semiotics is also relevant here. See especially Fernández 2016.

[5] See for example, Kiers and West 2015.

[6] For a critique of traditional information theory and a functional alternative, see Corning and Kline 1998a,b; also, Corning, 2007b. For a perspective on the synergistic nature of human language and its integration with other cognitive abilities, see Szathmáry and Számadó, 2008.

[7] For an up-to-date overview, see the fiftieth anniversary issue of the *Journal of Theoretical Biology* in 2012 with some 20 articles devoted to various aspects of cooperation. For a critical review of the work on sanctions, see Frederickson 2013.

[8] (Some of the many contributions on this theme are cited in an outtake at my website.)

[9] Corning and Szathmáry 2015; Szathmáry 2015.

[10] See especially Calcott and Sterelny, eds. 2011.

[11] Corning and Szathmáry 2015. (See the summary in the outtake at my website.)

[12] McShea and Simpson 2011.

[13] For the record, I differ slightly from Maynard Smith and Szathmáry, who use a dualistic definition that includes new forms of information (see below). I also agree with Ereshefsky and Pedroso (2015) that a new level of reproductive specialization need not be involved (e.g. humans). (See Chapters 7 and 8.)

[14] The term has many fathers. Among others, see especially Buss 1987; Ghiselin 1997; Michod 2007; also, West *et al.* 2015; Ereshefsky and Pedroso 2015.

[15] See Heckman *et al.* 2001; also, Field *et al.* 2015.

[16] Schrödinger 1944.

[17] For a critique of misuses of the Second Law, see Corning and Kline 1998a.

[18] Schrödinger's characterization is seductive. It has been cited on innumerable occasions over the years. But there are problems. (See the critique in the outtake at my website and Corning and Kline 1998a,b.)

[19] Corning and Kline 1998a,b; Corning 2007b. See also the brief history and discussion of the science of cybernetics in Capra and Luisi 2014.

[20] Maturana and Varela 1980/1973. As Capra and Luisi (2014, p. 132) put it, "Life is a factory that makes itself from within."

[21] For an insightful discussion of autopoiesis, see Capra and Luisi 2014. For a brief review of how the concept of autopoiesis is used, and often misused, see http://en.wikipedia.org/wiki/Autopoiesis (last modified 30 March 2015).

[22] See Kauffman 2008. Needless to say, the issue of how to define life has become a theoretical quagmire. A recent attempt to do so, using E-coli bacteria as a "model", can be found at http://www.ncbi.nlm.nih.gov/pubmed/25796394 See also the pioneering theoretical work by Robert Rosen 1991.

[23] Gánti 2003/1971; 2003. A more recent, independently developed formulation of this idea is the so-called CMP model, which stands for Container, Metabolism, and Program. See Bedau 2012.

[24] See especially Morowitz 1992. The theory developed by David Deamer, Harold Morowitz and others postulates a key role for what they call "amphiphiles" – elongated fatty molecules that are like phospholipids in modern cells. See especially Deamer ed. 1978, 2011; Deamer and Oro 1980; Morowitz 1978, 1981, 1992; Morowitz *et al.* 1987. (For more, see the outtake at my website.)

[25] Kauffman 1971, 1986; Hordijk *et al.* 2011; Vasas *et al.* 2012. See also Van Segbroeck *et al.* 2009; Bedau 2012. Another autocatalytic model, called the "hypercycle", was proposed by Manfred Eigen in the early 1970s. See especially Eigen and Schuster 1977, 1979. Maynard Smith and Szathmáry later observed in *The Major Transitions* that this model was unstable due to its susceptibility to "cheating". Szathmáry (2015) points out that many theorists confuse the hypercycle and chemoton models. They are not the same thing.

[26] See Corning 2005, 2007b, 2014; see also Mayr 1974. Jablonka (2002) develops a functional definition of information, which is similar to my concept of cybernetic "control information" and also implicitly recognizes its teleonomic foundation. Capra and Luisi (2014) formulate the same dynamic in terms of the concepts of autopoiesis and "cognition".

[27] The much-debated challenge of how to define life has been inconclusive. I take the contrarian view that life is not a discreet, static set of traits or attributes but a moving

target – a dynamic process that has been evolving (and changing) for some 3.8 billion years – so far.

[28] An excellent summary of this concept and the research behind it can be found at: http://en.wikipedia.org/wiki/RNA_world (last modified 8 May 2015).

[29] Indeed, the vital role of DNA in biosynthesis is possible only because of the highly coordinated role that three distinct forms of RNA play in "transcribing" and utilizing the nucleotide bases. See Darnell *et al.* 1990. (For more details, see the outtake at my website.)

[30] Szathmáry 2015. Others vehemently disagree (e.g. Capra and Luisi 2014). Some of the controversy involves different definitions of what the term means. For an up-to-date review and assessment, see Kun *et al.* 2015.

[31] Mazur 2014.

[32] Miller 1953; Miller and Urey 1959.

[33] Cairns-Smith 1987/1982; 2009/1968.

[34] However, the idea was recently revived with a new proposal, backed by some experimental work, that ancient clay based "hydrogels" could have served as an alternative to lipids as containment vessels. See Yang *et al.* 2013; also, see http://www.natureworldnews.com/articles/4784/20131106/life-evolved-clay-researchers-find.htm

[35] Wickramsinghe 1974.

[36] See especially Callahan *et al.* 2011. See http://en.wikipedia.org/wiki/Panspermia (last modified 27 May 2015). See also Crick 1981; Hoyle 1983.

[37] Wächtershäuser 1988, 1990, 1992.

[38] See Tunnicliffe 1991.

[39] See especially Gold 1992, 1999; Martin and Russell 2003.

[40] Wächtershäuser 2006, 2007; also, Wächtershäuser and Adams 1998.

[41] Russell 2006; Martin and Russell 2003, 2007; also, Koonin and Martin 2005.

[42] Lane 2009.

[43] Martin and Russell 2003, 2007; Martin *et al.* 2008; Lane 2009, 2014.

[44] Lane 2009. (For more details, see the outtake at my website.)

[45] Patel *et al.* 2015; also, Powner *et al.* 2009; Lilley and Sutherland 2011.

[46] Letter to J.D. Hooker, 1 February 1871. See Francis Darwin 1887, Vol. 3, p. 18.

[47] Wolfenden *et al.* 2015; Carter and Wolfenden 2015.

[48] Carter quoted in http://phys.org/news/2015-06-evidence-emerges-life.html. (accessed 6 June 2015). David Deamer (2011) disagrees with the hot pond scenario, citing experiments in which clays and metal ions blocked chemical interactions. On the otherhand, William Martin and his co-workers, in a pioneering DNA analysis (Weiss *et al.* 2016), believe they can trace the earliest common ancestor of contemporary living organisms to deep sea vents some 4 billion years ago. More research is obviously needed.

[49] With a somewhat different model, Capra and Luisi (2014, p. 142) nevertheless agree: "Life is the synergy of the three domains." Likewise, in a major review of the case for an RNA world, physicist Paul Higgs and chemist Niles Lehman (2015) conclude that "cooperation at the molecular level is essential" for the emergence and evolution of life. (See the outtake on this at my website.)
[50] Equally important, even at the most basic biochemical level, life is fundamentally a synergistic effect. (See the outtake at my website.)
[51] The following discussion is derived from Corning 2003, 2005; see also the reviews and references in http://en.wikipedia.org/wiki/Bacteria (last modified 2 May 2015), and http://en.wikipedia.org/wiki/archaea (last modified 15 May 2015).
[52] Christina Warinner, Prehistoric Human Biology as Inferred from Dental Calculus, in the 2016 CARTA public symposium on Ancient DNA and Human Evolution, https://carta.anthropogeny.org/node/156412 (accessed 29 May 2016). Sears 2005.
[53] See Margulis and Sagan 2002.
[54] Price 1991.
[55] Gorbach 1990; O'Hara and Shanahan 2006; Zoetendal *et al.* 2006.
[56] See the review in Crespi 2001; also Allison *et al.* 2001. A useful analysis can be found in West *et al.* 2006.
[57] (For some issues related to stromatolites, see the outtake at my website.)
[58] Bonner 1988. See also the discussions of collective behavior in bacteria by Shapiro 1988; also, Shapiro and Dworkin, eds. 1997.
[59] See http://en.wikipedia.org/wiki/Bacteria (last modified 2 May 2015).
[60] The following discussion is an updated version of Corning 2003, pp. 57-59.
[61] The case for this is argued in some detail in Lane and Martin 2010. See also Lane 2014. (For more, see the outtake at my website.)
[62] The combination of mitochondria and chloroplasts in plants enables them to generate some 15-20 times as much available energy (net of entropy) as do prokaryotes. See Margulis and Sagan 1995.
[63] Biologist Thomas Cavalier-Smith has pointed out that phagocytosis in turn depended on the prior development of internal membranes and the cytoskeleton. This allowed for the loss of a rigid outer cell wall in favor of a flexible external membrane that could create pockets. Cavalier-Smith 1981, 1991, 2013.
[64] Mereschkovsky 1909.
[65] Quoted in Khakhina 1992a; Khakhina 1992b; Margulis & McMenamin, eds. 1993.
[66] Wallin 1927, p. 8.
[67] See especially Margulis 1970, 1981.
[68] See especially de Duve 1996.
[69] Reviewed in Margulis and McMenamin 1990; also, Margulis and Dolan 1998.
[70] See Margulis *et al.* 2000.
[71] Reviewed in Szathmáry 2015. See also Cavalier-Smith 2009, 2013.
[72] Martin and Müller 1998. See also Vogel 1998.

[73] Martin and Mentel 2010. (For another option, see the outtake at my website.)

[74] Lane 2009, pp. 111-112. See also Lane 2014.

[75] Queller 1997, 2000; also, Queller 2004.

[76] Bell 1985; Michod 1999, 2007, 2011; Michod and Herron 2006.

[77] Bonner 2006; also, Bonner 1999. Of course, large size can be a severe disadvantage if the food supply is short, and bacteria manage to do alright despite their small size.

[78] Halstead 1988.

[79] http://en.wikipedia.org/wiki/List_of_muscles_of_the_human_body (last modified 5 June 2015).

[80] http://en.wikipedia.org/wiki/List_of_organs_of_the_human_body (last modified 21 May 2015).

[81] This supplants an earlier estimate of 51 areas in the so-called Brodman Scheme. See https://en.wikipedia.org/wiki/Human_brain#Functional_divisions (last modified 8 June 2015). http://en.wikipedia.org/wiki/List_of_animals_by_number_of_neurons (last modified 9 June 2015). Also, see Glasser *et al.* 2016.

[82] Szathmáry 2015.

[83] Jablonka and Lamb 2006; see also Ginsburg and Jablonka 2010.

[84] Tagkopoulos *et al.* 2008; Mitchell *et al.* 2009.

[85] Szathmáry (2015) defends the inclusion of humankind on the list of transitions based on what he calls our four distinctive traits: (1) language, (2) human cooperation, (3) eusociality (meaning a non-reproductive caste, such as grandparents), and (4) cultural group selection. (For more on this issue, see the outtake at my website.)

[86] Economist Paul Seabright, in his 2004 book, *A Company of Strangers,* explores in depth the phenomenon of widespread cooperation among non-kin as a hallmark of our species, along with exchange and markets) as a key institutional development and trust as a key lubricant.

[87] The meaning of this term is much debated. Some theorists apply it only to humankind (my preference as well) while others equate it with the concept of a superorganism and include any species that has achieved a complex social organization, such as leaf cutter ants. See especially Turchin 2013, 2016; Gowdy and Krall 2014, 2015.

[88] D.S. Wilson 2015, p. 49. See also the discussion of multi-level selection in Okasha 2005.

Chapter 7

The Self-Made Man I: Australopithecine Entrepreneurs

How can we account for the trend that has resulted over time in the remarkable evolutionary transformation of our lineage from arboreal apes to complex technological civilizations? Was it blind luck – a meandering "drunkard's walk" (to borrow a term from Stephen Jay Gould) with no particular logic to it? Or was there some underlying principle? I believe that there was an inner logic to the process, and that synergy, and Synergistic Selection, were deeply involved in shaping the trajectory of our evolution.

Human evolution is a notorious theoretical minefield, for the simple reason that the ratio of guesswork to evidence has always been very high. The well-known science writer Nicholas Wade points out in a recent book that the vast majority of evidence about our origins is lost forever. Wade laments, with some exaggeration, that the only hard evidence we have found so far for 90 percent of our history is "a handful of battered skulls and a few stone tools."[1] (Actually, an array of new research tools – from DNA analyses to the study of dental plaque in fossil teeth – have recently augmented our research arsenal.)[2] It's not surprising that just about every new fossil find seems to change the theoretical landscape, or that the scholarly debate about our "origins story" is often polarized and sometimes heated.

Here I will try to step back from the *mêlée* a bit and advance a scenario that is grounded in a plausibility argument and some overlooked logic about what was most likely to have occurred in human evolution, while also trying to be consistent with the available evidence. Two themes stand out in this scenario. One is the decisive role of various kinds of economic synergy (and Synergistic Selection) as the drivers of the process. The other major theme is the compelling evidence (to me and some others) that human evolution is a major example of a Lamarckian dynamic, properly understood (see Chapter 5).

Behavioral "Pacemakers"

In a nutshell, there is reason to believe that behavioral innovations and cultural changes by our remote ancestors were the "pacemakers" at every stage in human evolution, and that the anatomical changes that occurred over time tracked these cultural innovations, rather than the other way around. In a very real sense, the human species invented itself.

In my 1983 book, *The Synergism Hypothesis,* where I first proposed this idea, I invoked the title of archaeologist V. Gordon Childe's classic 1936 study of the agricultural revolution, *Man Makes Himself.*[3] A similar thesis was later developed independently by anthropologist-ecologist Robert Boyd and ecologist Peter Richerson, and by the zoologist Jonathan Kingdon in his insightful 1993 book with the clever title, *Self-Made Man.* Most recently, anthropologist Joseph Henrich has made a comprehensive case for this scenario in his engaging and readable book, *The Secret of Our Success.*[4]

This paradigm could also be classified as a gene-culture co-evolution theory (in the current parlance), or, perhaps better said, a culture-gene theory.[5] It also resonates with the important new field in biology (mentioned earlier) called niche construction theory, which is concerned with how living organisms can modify their environments and thereby affect the course of natural selection.[6] It is becoming increasingly clear that behavioral and biological changes have been deeply intertwined in human evolution, although behavioral innovations may have led the way. With a nod to Jonathan Kingdon, I will call it the Self-Made Man scenario.

Many disciplines are involved these days in trying to decipher our evolutionary origins – paleoanthropology, archaeology, primatology, human genetics, molecular biology, paleoecology, behavioral ecology, climatology, and more. Over the years, a number of different scenarios have been developed from the gradually accumulating body of evidence – including painstakingly discovered hominin fossils, DNA studies, indirect data about ancient climates, comparisons (and contrasts) with other species, plus a large dose of inference and even speculation. However, there is as yet no definitive version of our Genesis story. Much still depends upon how this assemblage of evidence is interpreted. The result has been a wealth of scenarios and a wealth of conflicting opinions. Science writer Roger Lewin, a diligent student of human evolution, wrote an entire book some years ago about the debates among the fossil-hunters with the provocative title *Bones of Contention*.[7]

Prime Mover Theories

One result of this cacophony is that there have been at least eight different "prime mover" theories of human evolution over the years featuring a supposed breakthrough, or a key step that shaped the process. These theories are detailed in an outtake at my website: www.complexsystems.org. Here I will provide only a brief summary.

The earliest, and for many decades the most frequently touted, theory of human evolution could be called the "warfare hypothesis," which traces back to Darwin's own speculations in *The Descent of Man.* Darwin's views about human evolution were balanced. He certainly understood the role of cooperation. (There is an outtake on Darwin's scenario at my website as well.) However, he also believed that competition and differential selection between evolving human groups was an important influence, especially in shaping our social and moral faculties. As Darwin wrote:

> Natural selection, arising from the competition of tribe with tribe…would, under favorable conditions, have sufficed to raise man to his high position…A tribe rich in the above [moral] qualities would spread and be victorious over other tribes; but in the course of time it would, judging from all past

> history, be in its turn overcome by some other tribe still more highly endowed.[8]

Many of the theorists who followed in Darwin's footsteps equated his "competition of tribe with tribe" with violent conflict or warfare and portrayed this dynamic as the predominant influence in shaping human evolution.[9] However, the "killer ape" scenario (as it has sometimes been called) was challenged in the 1960s by an alternative theory that came to be known as "Man the Hunter" – after the title of a landmark conference and an edited conference volume that followed. In the introduction to this book, Sherwood ("Sherry") Washburn and Chet Lancaster – two of the leading anthropologists of the day – characterized big game hunting with weapons as the "master behavior pattern" of our species.[10] They viewed warfare as a later outgrowth of this pivotal social development.

Man-the-hunter was a popular idea at first, but it was soon opposed by an alternative prime mover theory dubbed "woman the gatherer."[11] The advocates for this scenario claimed that meat had played a relatively small role in the evolving hominin diet compared to fruits, nuts, leaves and other gathered foods, and that the females were the primary providers for our remote ancestors.[12] It was, needless to say, a very controversial theory.

The Food Sharing Hypothesis

A fourth prime mover scenario, viewed by many theorists at the time as a kind of middle-ground between the two supposedly "sexist" models, was the "food sharing hypothesis" proposed in the late 1970s by the anthropologist Glynn Isaac.[13] Isaac pointed out that food acquisition is typically a "corporate responsibility" in modern hunter-gatherer societies, in contrast with chimpanzees and the other great apes. He posited that a sharing of gathered plant foods and hunted (and butchered) meat at designated home bases was a key innovation in human evolution. While Isaac's theory was attractive, it was subsequently battered by various critical analyses.[14]

A more elaborate argument for food sharing and male provisioning, one that did not rely on hunting/scavenging, was developed by Owen Lovejoy in the early 1980s. His scenario is often referred to as the "nuclear family model."[15] Lovejoy's reasoning was based on what he perceived to be a major "demographic dilemma" (as he called it) for our remote Pliocene ancestors (from 5.3 to 2.5 million years ago). While the hominins of that era produced relatively few offspring, each one of them required a relatively high level of "parental investment," both for prenatal development and for postnatal infant care. How could they be properly fed? The obvious solution to this dilemma, Lovejoy believed, was to recruit the males to provide food for the mothers and their infants – as many birds and most canids do – along with shifting to a monogamous mating pattern. Lovejoy's scenario was appealing, but it too was sharply criticized.[16] Such a radical change seemed implausible.

Cooking, Climate, and Confrontation

Three more recently developed theories, sometimes also treated as prime movers, should also be mentioned. One is the so-called "cooking hypothesis."[17] Anthropologist Richard Wrangham and his colleagues believe this cultural development was even more important than the addition of meat to our ancestors' diet and that it played a key role in our evolution. (We will return to this scenario in the next chapter.)

Another recent theory is focused on the growing evidence that our remote ancestors had to cope with a climate, and an environment, that was highly variable over time.[18] For instance, we now know that there has been a total of 27 ice ages during the past 3.5 million years alone, and that many of these episodes involved major changes in land forms and local ecosystems in various parts of the world.

Some theorists believe that these environmental changes were sufficient to account for the origin and evolution of the hominin line. They are said to have "induced" or "driven" or "forced" East African apes to shift from an arboreal lifestyle to terrestrial living and to adopt new survival strategies. As one well-known theorist put it, "the causes were ecological."[19] The evolutionary biologist Patrick Bateson disparages this idea, calling it the "billiard ball" theory of human evolution.[20] (More on this matter below.)

Finally, there is the much-debated theory that attributes the emergence of *Homo erectus* and our big brains (and language) to the confrontational scavenging" of carrion from dead animals, in competition with other East African carnivores.[21] In fact, it was the legendary fossil hunter Louis Leakey who first proposed, in the 1960s, that a scavenging niche might have preceded the evolution of hunting by our early ancestors. Other theorists have made similar suggestions, and the idea was picked up and explored in depth in the 1980s and 1990s by anthropologist Robert Blumenschine and his colleagues.[22] The "power scavenging" theory (as it is also sometimes called) has since descended into a rancorous debate about different models, different interpretations of the existing data, and the meaning of new findings.[23]

In retrospect, the common denominator in all eight of these prime mover scenarios is that they invoke behavioral changes, not genetic mutations or anatomical alterations, as the drivers, just as Lamarck himself might have proposed. Even the climate change hypothesis posits that major environmental shifts induced behavioral responses. It's very possible that every one of these scenarios has some merit. As we shall see, each of them may have played some role at different times and places, although they were certainly not the whole story.

The Self-Made Man Scenario

In contrast with these prime mover theories, the Self-Made Man scenario proposes that our remote ancestors gradually assembled a unique synergistic package – a remarkable suite of traits that, in combination, enabled our species to gain increasing control of its destiny to a degree that no predecessor has been able to do. The human narrative was shaped by an accumulation of various forms of synergy with economic advantages for the problems of survival and reproduction. The process was propelled by a series of synergistic behavioral innovations, and by a proximate dynamic of Synergistic Selection (cultural selection).

In the truest sense, the evolution of humankind involved an entrepreneurial process – a pattern of invention, trial-and-error learning, selective retention, and cultural transmission, which in turn shaped the subsequent evolution of supportive anatomical changes. Functional synergy played a key role in this transformation. It resulted in a variety of bioeconomic benefits – synergies of scale, threshold effects, new combinations of labor, functional complementarities, cost- and risk-sharing, and more – all of which were highly advantageous for survival and reproduction.

Before proceeding with this scenario, however, two preliminary points are in order. One is that the basic assumptions that are made about the context of human evolution are critically important. In many of the earlier prime mover theories, the full context of human evolution was unstated, or downplayed, or used only selectively to develop a hypothesis. The problem is that biased assumptions can lead to logically-impeccable conclusions that are totally wrong.

The other preliminary point is that, in speculating about the many unknowns in human evolution, the principle of parsimony (Occam's razor) provides a useful analytical tool – a way of choosing among various explanations when hard evidence is limited. To be sure, Occam's razor has a double-edged blade, so it must be used with care. But, all other things being equal, it seems more likely that the simplest and most incremental (least radical) alternative will be the right one. By the same token, a scenario that involves the highest rewards or payoffs for the least possible cost or risk (cost-effectiveness) is more likely to have been the one that our ancestors selected, by and large. Although it's unlikely that we will ever find conclusive evidence for many aspects of human evolution, we may be able to rank-order the different alternative options in terms of their relative merits and plausibility.

Scoping Out the Context

The first and perhaps most fundamental point about the context of human evolution is that, as noted earlier, our ancestors had to cope with an environment that was highly variable over time.[24] For instance, during the deep freeze on Earth that began about 33,000 years ago and peaked about 13,000 years later, sea levels declined more than 400 feet from present levels as a mile-high mountain of ice became trapped in the expanding northern glaciers. Then the process went into reverse. Similar climate swings can be traced back millions of years.

The anthropologist Richard Potts has long argued that the challenges associated with extreme climate variability were a major shaping influence in human evolution.[25] However, firm evidence of a close correlation between climate changes and any specific benchmark in our evolution is sparse and contradictory – and much debated. An analysis of the extinction pattern in 510 mammal species in the Lake Turkana region in the 1990s failed to support this linkage.[26] In another study, anthropologist Tim White concluded that "The strongest pulse of speciation and extinctions in both pigs and hominids... occurs between ca. 1.6 and 2.0 [million years] ago during a period in which no dramatic global climatic oscillation signal is available."[27] On the other hand, more recent data and analyses by climatologist Peter de Menocal suggests that some *local* climate changes do correlate with hominin evolution, especially the extinction of the australopithecines and the emergence of *Homo erectus* (and their relatives).[28]

A Moving Target

One problem with treating these correlations as causal linkages is that new fossil finds keep changing the time-course of human evolution; it's a moving target. But more important, a model that treats climate as a deterministic causal agency obscures the active role of the participants themselves. Our ancestors were not "billiard balls" (after Patrick Bateson). A gradually shifting resource base may have changed the "payoff matrix" – the relative abundance and the costs and benefits of exploiting various alternative food sources, as well as the intensity of competition for these resources. Likewise, a mosaic of forested and open areas may have created many new ecological opportunities.

But exploitation of this new environment required initiative and a "learning curve," along with mobility. The ability to safely traverse open areas and (most likely) to defend resource patches against competitors was critically important. Some of the East African primates of that time chose to change their "manner of procuring subsistence," in Darwin's phrase. Others did not. We are descended from these pioneers.

The shift to a terrestrial life style most likely did not happen all at once. For one thing, it involved substantial costs and risks. As foraging ranges expanded, so did the time and energy needed to exploit various resources. However, the australopithecines (and especially their *Ardipithecus* predecessors) were imperfect bipeds – competent but not as efficient as later *Homo erectus* (also sometimes called *Homo ergaster*).[29] More important, the exploitation of a terrestrial environment introduced new risks from predators and competitor species, not to mention rival hominin groups. Some theories of human evolution have downplayed these threats, but it was in fact a major challenge, with life and death consequences.[30]

It is possible for a small primate to survive for many days without food, so long as fresh water is available, but a single encounter with a major predator will very likely reduce its survival chances to zero. There is fossil evidence going back at least to *Ardipithecus ramidus* some 4.4 million years ago suggesting that our remote ancestors were indeed subject to severe "predation pressure," as the ecologists would say. In those days, there were no less than ten large carnivore species roaming East Africa, compared with just four today. Pack-hunting species like *Palhyaena* would have been particularly dangerous.[31]

Primate Pre-adaptations

The primates that chose to venture into this changing and hazardous environment brought with them many important precursors/pre-adaptations (or "exaptations" if you prefer this fashionable neologism) – stereoscopic color vision, a modified climbing anatomy, dexterous forelimbs with manipulative hands, sociality, a relatively high degree of intelligence, an omnivorous dietary pattern, and – very likely – a form of social organization that is based on a nucleus of closely-related males who are joined by unrelated females. Very few primate species have male-based groups, and this may have been an important enabler for the emergence of the hominin adaptive pattern.[32] In other words, the initial phase of our evolution may have followed David Queller's "fraternal" pathway (see Chapter 6).

Two other precursors might also have been important. One had to do with the fact that our earliest hominin ancestors were relatively small – three to four feet tall. Even the much later *Australopithecus afarensis* of 3.5 million years ago are estimated to have weighed some 15 percent less than modern chimpanzees. Anthropologist Milford Wolpoff has called them "Miocene midgets."[33] (The Miocene spanned the period from 23 to 5.3 million years ago.)

Was Bipedalism a Pre-adaptation?

A second point is that – contrary to an assumption going back to Darwin himself – bipedalism may actually have been a pre-adaptation and facilitator, rather than a product of our evolutionary transition. The traditional view is that our ancestors evolved directly from "brachiators" that used their arms to move through the trees. However, a number of theorists have espoused the alternative idea that hominins evolved from an above-branch biped – rather like a tight-rope walker – that resembled a 10 million-year-old Sardinian primate called *Oreopithecus bambolii*.[34]

Bipedalism, in other words, may have been an ancient form of arboreal locomotion that was applied to a new situation (a new "habit") and was further improved upon over time via natural selection. Indeed, more recent research among contemporary Sumatran orangutans lends support to this theory.[35] Primatologist Frans de Waal also points out that bonobos – the so-called pygmy chimpanzees – use a bipedal gait more often than chimpanzees, especially when carrying objects.

Another intriguing hypothesis about our bipedalism was advanced by Jonathan Kingdon in his 2003 book *Lowly Origin* – a title he borrowed from an expression used by Darwin in *The Descent of Man* to characterize human evolution.[36] Kingdon pointed out that the numerous anatomical changes involved in becoming a bipedal ape, including four major modifications to the skeleton and trunk alone, must have been a piecemeal evolutionary process, with multiple steps. Furthermore, Kingdon argued, these changes were necessary pre-adaptations for being able to achieve a balanced, upright, two-legged gait. He proposed that these anatomical changes were caused by the adoption of a terrestrial "squat feeding" pattern that included both extensive ground foraging and tree climbing for access to fruits, and for safety.[37]

Of course, the *Oreopithecus* and squat feeding models are in conflict with the widely accepted view that contemporary chimpanzees represent the most likely starting point for our own evolutionary divergence. Our common ancestor is often presumed to have been a knuckle-walking fruit eater that was similar to modern chimps. Some years ago, Richard Wrangham developed a list of chimpanzee traits that he supposed would also have existed in our common ancestor,[38] while Jared Diamond wrote an entire book based on this assumption called *The Third Chimpanzee.*[39] The problem is that we really do not know which model is more likely to be correct, because we have not found the common ancestor. In fact, we have not found any ancient chimpanzee fossils at all, so far as I know. There is also a dearth of hominin fossils dating back to the most critical time period, from 7–10 million years ago.

Bipedalism is More Energy Efficient

However, some recent research casts doubt on the idea that our remote ancestors were in fact knuckle-walkers.[40] The new research has shown that bipedalism is far more energy efficient than walking on all fours. For instance, modern humans consume only one-quarter of the calories used by chimpanzees for locomotion.[41]

Equally important, there are some major advantages to bipedalism. The most obvious advantage is that it frees up the hands to carry objects and engage in other useful activities, just as Darwin supposed. It would also have allowed ancient hominins to expand their day ranges and, over time, engage in long-distance foraging and even running in pursuit of game.

Whatever may have been the case, we now have strong evidence that bipedalism was a crucial factor in our successful transition to living on the ground and that it may date back to our earliest known hominin predecessors, the 6 million-year-old *Orrorin tugenensis* and even *Sahelanthropus tchadensis* of 7.2 million years ago.[42] Bipedalism was a foundational hominin trait.

Synergy in Australopiths

However, all of these pre-adaptations amounted only to facilitators. The crucial behavioral invention among our remote ancestors – the pacemaker for the path that we have taken – was the adoption of a group-based, cooperative (synergistic) survival strategy.[43]

The key question is, how did a diminutive ape with constrained mobility on the ground and no natural defensive weapons, but with a relatively large brain (on average about the size of a modern chimpanzee), manipulative hands, and an omnivorous digestive system solve the problem of shifting to a terrestrial habitat, broadening its resource base and, over time, greatly expanding its range? Social organization – group living – was a key factor. In a patchy but relatively abundant woodland environment that was also replete with predators and competitor species, group foraging and collective defense/offense was the most cost-effective strategy – indeed, the only viable strategy. A terrestrial food quest necessitated cooperation. As anthropologist Joan Silk puts it, there was "safety in numbers."[44] Equally important, there were immediate economic payoffs (synergies) for collective action that did not have to await the workings of natural selection.[45] (An analogy can be seen in the tightly organized and aggressive troops of modern baboons, the most successful of all terrestrial primates next to humans.)[46]

However, it is also likely that the earliest of these hominin pioneers stayed close to the safety of the trees. Early australopiths (and their precursors) apparently had very powerful arm muscles, perhaps to compensate for the reduction in climbing ability with their hind limbs.[47] Wolpoff compares them to modern-day steel workers.

Public Goods

It should be stressed that the group-defense scenario assumes only that the synergies derived from acting collectively – foraging and reproducing as a group – were both immediate and mutually beneficial (Occam's razor). The odds of survival were greatly enhanced. There may well have been group selection for this social activity, but it was not based on altruism. It involved "collective goods," or "public goods."[48] Because these groups were formed around a nucleus of closely-related males, individual selection, kin selection, and group selection would have been aligned and mutually reinforcing – just as Darwin proposed in *The Descent of Man*.

Why would the males defend the females and infants? First, the males might not have known their paternity if the females followed a reproductive strategy of multiple matings, as chimpanzees typically do. A second factor was that all of the infants would likely have been closely-related – "nephews," "cousins," or even younger siblings. A third factor is that, in a so-called "K-selected" species (one with relatively few offspring, a very long reproductive cycle, and a relatively short life span), each offspring was more valuable. The benefits of protecting the mothers and infants in a dangerous environment, and the costs of not doing so, were much greater. Finally, it was not much more costly to defend the offspring of close kin than it was to defend one's own children (even assuming that they could be identified). It was not a matter of altruism, or reciprocal altruism but of teamwork in a win-win (or lose-lose) situation. Again, a modern analogy can be found in terrestrial baboons, where even unrelated males cooperate in troop defense.[49]

A Division/Combination of Labor

Modern human societies typically exhibit a division of labor along sexual lines, and it is likely that a rudimentary version of this pattern also existed among the australopiths. Applying the principle of parsimony once again, it seems likely that the females would have been responsible for carrying and feeding the infants and shepherding the juveniles, while the males served as the primary guardians for the group. (Again, an analogy can be found in baboon troops.)

Some evidence for this scenario can be seen in the australopithecine fossils of 3–3.5 million years ago, where there appears to have been a sharp sexual difference in size. The females remained very small while the males grew much larger, closer in size to modern pygmies. Many theorists have attributed this size-difference to sexual competition among the males. It is well-known that sexual selection can produce significant anatomical differences between the two sexes. However, a division of labor between males and females can also create an advantage for dimorphism. If australopithecine males were close kin, this might have reduced status rivalries and male sexual competition. On the other hand, if the males came to play an important role in defending the group, larger size would have provided a significant benefit.[50] An analogy can be found (once again) in baboons, where the females average about 30 pounds and the males about 90 pounds, a much greater difference than is typically found in primates.

We may never know for certain about this and many other details related to human evolution, but group foraging and a cooperative division (combination) of labor that allowed for an increased access to a resource-rich but more dangerous environment seems likely to have been a key development in the hominin line. It would have involved the most limited, incremental behavioral changes with the most cost-effective payoffs, and it was highly synergistic. Moreover, as time went on, the group-living mode of adaptation encouraged other forms of social cooperation and more elaborate forms of synergy (more on this below). Culture, after all, is a collective phenomenon that is shared across generations, not an individual trait. Indeed, some theorists believe that the evolution of a more egalitarian social structure based on consensual leadership and group competence – another distinctive hominin adaptation – would have become highly advantageous even at this early stage in our evolution.[51] (More on this in Chapter 8.)

Tools and Weapons?

However, there is also a potential problem with this model. If high quality food resources are not always abundant, a large foraging group might become subject to disruptive internal "scrambling competition," as it's called. Other social primates, such as chimpanzees, solve this problem by splitting up into small foraging units during the day.[52] But modern-day chimpanzees don't have intense pressure from predators.

Accordingly, a second key to the australopiths' adaptive strategy may well have been the invention and use of tools and defensive weapons – a synergistic soft technology of wood and bone implements, and perhaps thrown objects as well.

There have been many tool-use advocates over the years, beginning with Darwin himself.[53] However, some theorists claim that the first real tools in hominin evolution were the stone cutters, choppers, and scrapers that date back some 2.5 million years (or perhaps even to 3.3 million years, according to new findings).[54] These theorists often downplay the importance of tools and envision the australopiths and their cousins as having been minimal tool-users at best. But this view is a bit obtuse. For one thing, the lack of fossil evidence is not evidence of a lack – as the old cliché goes. But more important, tool-use can have transformative consequences. It can open up a whole new ecological niche or facilitate a new adaptive strategy, as the niche construction theorists have highlighted. Previously unattainable sources of food can become a reliable, even an abundant part of an animal's food supply.[55]

Digging Sticks?

The current thinking is that the so-called underground storage organs (tubers, bulbs, rhizomes, corms, etc.) may have played an important role in the evolution of the australopithecines, because they provided a reliable, year-round source of food in a more seasonal woodland habitat.[56] However, this dietary shift also presented a major problem. How could these resources be exploited efficiently? Digging sticks may have provided the solution.

It seems very unlikely that the australopiths could have adapted successfully to a terrestrial life-style and survived for perhaps three million years without the skilled use of various natural objects as tools – digging sticks, stone hammers, cutting tools, food carriers and the like. Indeed, chimpanzees and even capuchin monkeys are frequent users of tools for procuring food, as noted earlier, and sometimes in conflict situations as well.[57] (We have recently learned that chimpanzees can even use crude spears to hunt other animals.)[58] It seems unlikely that the stone tools dating back as far as 3.3 million years ago sprang from the mind of an otherwise tool-challenged biped. It's more likely that the australopiths were prolific tool users.

A more distasteful proposition, though equally plausible I believe, is that weapons also played an indispensable part in the successful transition of the australopithecines to a terrestrial life-style. One can hardly exaggerate the value to a diminutive, relatively slow-moving biped of being able to use a short stick, or a sturdy femur, or even a well-aimed rock as a weapon.[59] One obvious advantage is that anyone wielding a weighted object can engage an opponent at much less personal risk (beyond arms' length) and can strike a much more damaging blow. Even crude weapons, if skillfully used, would have served as "equalizers" that changed the balance of power. Indeed, the same hand-held wooden implements might have doubled as digging sticks and weapons.

Here, once again, the principle of parsimony can be invoked. Consider the alternatives. A group of cooperating Miocene/Pliocene midgets without the benefit of defensive implements of any sort would have been at a great disadvantage against bigger, faster opponents that were endowed with large canines and claws. By the same token, a lone australopith, even armed with a weighty femur, would have been at a severe disadvantage against a pack of hungry hyenas. But a group of australopiths traveling together in dangerous or unfamiliar country armed with weapons and ready to act collectively (a synergistic combination) would have been far more likely to hold their own in any life-and-death situation. Even if these creatures were only infrequently subject to predation, just one incident in a lifetime would have been one too many.

A Pliocene Arms Race?

Although more speculative, it is also possible that the arms race in humankind began in the Pliocene. If groups wielding weapons were able to gain leverage in confronting predators, the same would have been true in confronting various competitors. Again, the principle of parsimony can be invoked. The fabrication and skilled use of tools/weapons by these early hominins were learned, socially-transmitted behavioral innovations, and their application to new contexts would have involved only an incremental step with potentially large benefits.[60]

Violent confrontations with other groups can be very dangerous, especially when weapons are involved. The costs could outweigh any potential benefits. It's more likely that armed group conflicts among the australopithecines were related to such vitally-important objectives as the acquisition or defense of food patches, water holes, and sleeping sites. Anthropologists Lisa Rose and Fiona Marshall call it the "resource defense" model, but there must (logically) have been an equal number of resource "offenses."[61] It's very possible that the male chimpanzee practice of raiding other groups to acquire females was also common among the australopiths, as various theorists have suggested.[62] In any case, organized collective violence, especially with crude weapons, represents another form of social synergy that, very likely, has deep roots in human evolution.[63] (More on this in Chapter 8.)

Midwifery?

One other form of social cooperation (and synergy) may also have played an important role in the evolution of the australopithecines. Human females have a much more difficult birthing process than, say, chimpanzees, and our larger brain is partly to blame. Less widely appreciated is the fact that changes in the anatomy of the pelvic region associated with the shift to bipedalism were also a complicating factor for our hominin ancestors. In chimpanzees, the infant's head emerges at birth facing forward. The mother can see the infant's face, can manually aid in the delivery process, and can take care of the afterbirth details (clearing the infant's nose and mouth, cutting the umbilical cord, etc.). However, the anatomical changes evident even in australopithicine females suggest that the human birthing process, with the infant facing the mother's back, had already become a necessity; the infant's head had to rotate in order to fit through a more constricted "pelvic aperture." One consequence was that the childbirth process in the australopiths may have required assistance – midwifery.[64]

A Social Triangle

So, in addition to male-male and male-female cooperation, cooperation between females may also have become one of the elements of the australopiths' adaptive package. (An analogue can be found in the high level of female cooperation in bonobos – the pygmy chimpanzees.)[65] In other words, there was a social triangle – a three-way nexus of cooperation and synergy. And this triad provided the scaffolding for the development of a more elaborate pattern of cooperation over time. Humankind evolved as a closely cooperating, group-living genus. And this, in turn, facilitated the development of many other social synergies as well.

The key to australopithecine sociality may have been the economic costs and benefits to each individual for cooperation or non-cooperation. Reciprocity and reciprocal altruism may help to explain it. But the benefits associated with being included in the group – and the high (even fatal) cost of being ostracized – may have been a major factor as well. Each individual had a stake in the viability and well-being of the group as a survival unit. The group represented a "common good" that provided collective synergies. And this would also have created an incentive for measures to contain conflict and enhance cooperation. For instance, a larger group was more likely – all other things being equal – to benefit from synergies of scale in confrontations with other groups of predators and competitors (not to mention potential prey). It seems likely that the economic principle of "public goods" – collective synergies – has undergirded human evolution from a very early date.[66]

There is some hard fossil evidence for this cooperative, group-living, tool-using hypothesis. One source comes from brain endocasts – impressions of the interior surface of australopith skulls made by anthropologists Ralph Holloway, Philip Tobias, and others.[67] These surface impressions suggest that, even 3.5 million years ago, australopithicine brains, while still quite small by comparison with modern humans, may already have undergone some internal reorganization. The changes were of the kind that are associated with skilled use of the hands, more complex social behaviors and more sophisticated communications abilities – though probably not language as we know it. (More on language in Chapter 8.)[68]

This scenario also accords with the "social brain" hypothesis, proposed by anthropologist Robin Dunbar.[69] His theory holds that there is a strong correlation, especially in primates, between brain size/ complexity and social group size and complexity. Psychologist Michael Tomasello, likewise, argues that our unique cognitive skills have been shaped by social interactions involving a "shared intentionality," as he calls it.[70]

The implication is that, during its first two million years or so, the australopithicines had moved well beyond the cognitive and social abilities of modern chimpanzees in ways that were consistent with the scenario described above; they were accomplished group-living, tool-using, communicative bipedal apes, and the changes in their brain anatomy were a reflection of major changes in their behavioral repertoire. They were opportunistic omnivores that foraged in closely cooperating, interdependent groups.

Learning from Nature

Jonathan Kingdon has pointed out that these creatures would also have had both the incentive and the ability to make more systematic use of the various materials and resources in their environment, and plenty of models to inspire them. Like many other species, our early ancestors must have learned from observing nature – spider webs, nest building techniques, food procurement strategies, the hunting tactics of large carnivores, and the like.[71] Thus, a chance observation by a foraging group that some other small mammal was digging up an edible tuber with its forepaws might have led to the deployment of digging sticks by the group to exploit this food source for themselves. Again, the rewards for adopting such behavioral innovations would have been immediate – Synergistic Selection at the proximate level.

A second form of hard evidence consists of changes in dentition – tooth size and shape, enamel thickness and other characteristics – which anthropologists rely upon to differentiate among the various fossil discoveries. Australopithecine dentition was significantly different from that of chimpanzees. The reduction in their canine teeth and the development of larger molars with thicker enamel suggest that they had adopted a more diversified dietary pattern, with a larger percentage of processed and chewed plant foods and meat. This, in turn, implies that behavioral changes preceded the natural selection of various anatomical changes.[72]

"Improved" Hands

A third source of evidence involves a significant alteration in the hands of the austalopithecines. These early hominins had relatively dexterous hands compared to other primates and the great apes. Their thumbs were shortened and their fingers were flattened, although the thumbs were still not fully opposable. The modern precision grip was certainly not perfected in australopiths, so it is unlikely that their tool-making was very advanced. However, anthropologists Nicholas Toth and Kathy Schick, who have studied early hominin tool-making techniques in depth, found that even the crude, "flaked" stone tools of 2.5–3.3 million years ago required great manual dexterity and considerable practice. Among other things, Toth and Schick showed that even an intelligent and carefully trained bonobo, the famous Kanzi, was unable to master these skills.[73]

Thus, the australopiths most likely could do things with their hands that chimpanzees could not – although they were far less adroit than their successors. Much has been made of the cognitive developments underlying the emergence of language, but the developments associated with skilled use of the hands were equally significant, as Darwin long ago supposed. In fact, it seems that these skills are neurologically linked to language skills and may have been a facilitator for language development as well.[74]

One further implication of the group foraging and group defense/ offense model is that the carrying of objects may have played an important part in australopithecine evolution after all – just as Isaac and Lovejoy suggested (as noted earlier) – but for different reasons. As our ancient hominin ancestors began to move into more exposed environments, the females had to carry the infants, and perhaps digging tools, to new foraging sites, while the males carried tools/weapons that could readily be used for defense.

At first, no doubt, these forays were very tentative; hominin groups probably remained close to the safety of the trees. But, as their skills, their size and their confidence increased, they were able to migrate over longer distances and occupy more diversified environments. As far back as 3 million years ago, australopiths had spread throughout Africa; fossil remains have been found as far as 2,000 miles away from their East African heartland. It also appears that they were thriving in much different climates with more varied seasonal resources. Intensive scavenging and small-game hunting may well have become an imperative for some of them.[75]

Thus, far from being a way station, or a stepping stone on the road to *Homo sapiens*, the australopithecines were a path-breaking and immensely successful genus that lasted for perhaps three million years. And then they disappeared. The most likely culprit, I believe, was *Homo erectus.*

---

[1] Wade 2006, p. 1. See also Bokma *et al.* 2012.

[2] Paleoecology has developed an array of sophisticated new tools in recent years, including stable isotope analyses, feces analyses, and ancient micro-climate analyses, allowing us to extend what little hard evidence we do have.

[3] Corning 1983; also 2003, 2005.

[4] Boyd and Richerson 1985, 2005; Richerson and Boyd 2005; Kingdon 1993; Henrich 2016. See also Fisher and Ridley 2013.

[5] Gene-culture co-evolution theory, also known as "dual inheritance theory," is attributed to three major contributions in the 1980s by Lumsden and Wilson 1981; Cavalli-Sforza and Feldman 1981; and Boyd and Richerson 1985. For an overview, see https://en.wikipedia.org/wiki/Dual_inheritance_theory (last modified 8 June 2105). See also Henrich 2016. He also prefers the "culture-gene" version.

[6] For the important work on "niche construction theory," see Odling-Smee *et al.* 1996, 2003, 2013; Laland *et al.* 1999; O'Brien and Laland 2012.

[7] Lewin 1997.

[8] Darwin 1874/1871, p.148.

[9] Perhaps most famous was Raymond Dart's (1953, 1959) portrayal of our ancestors as blood-thirsty and ravenous "killer apes." (For more, see the outtake at my website.)
[10] Washburn and Lancaster 1968, reprinted in Ciochon and Fleagle 1993, p. 219. (Some subsequent misinterpretation is noted in an outtake at my website.)
[11] Sally Linton's (1971) seminal paper appeared years earlier, but the debate was ignited by Adrienne Zihlman and Nancy Tanner (1978) in their provocative paper.
[12] In the Zihlman and Tanner female gathering scenario, the males were also viewed as being economically superfluous. (For details, see the outtake at my website.)
[13] Isaac 1978. See also Isaac 1981, 1983.
[14] Binford 1987; Potts and Shipman 1981; Potts 1984, 1988; Shipman 1983, 1986.
[15] Lovejoy 1981; see also Lovejoy 2009.
[16] (See the outtake on Lovejoy's scenario at my website.) In contrast, see the work on "allomothering" by Hrdy 2000, 2009; Hawkes 2003; also, Kramer and Otárola-Castillo 2015. More on this in Chapter 8.
[17] Wrangham *et al.* 1999; Wrangham 2009; Wrangham and Carmody 2010; Gowlett and Wrangham 2013; also, Pennisi 1999.
[18] See the various contributions in Vrba *et al.* eds. 1995; also, Kingdon 1993; Foley 1994,1995; Allen *et al.* 1999; McManus *et al.* 1999; Taylor 1999; Alley 2000; Keeling and Whorf 2000; Stanley 2000; Richerson *et al.* 2001; deMenocal 2004, 2011; Richerson and Boyd 2013; Potts and Faith 2015.
[19] Vrba *et al.* eds. (1995) proposed that climate changes were the necessary and sufficient cause for the emergence of *Homo erectus*. According to their "pulse hypothesis," climate changes have been the primary drivers for various waves of speciation and extinction. See also Stanley 1992.
[20] Bateson and Hinde, eds. 1976; Bateson 1988. See also Foley 1995.
[21] See Washburn and Lancaster 1968; Blumenschine 1987; Shipman and Walker 1989; Wrangham and Peterson 1996; Stanford 1999; Wolpoff 1999a; Wood and Collard 1999.
[22] See Blumenshine 1987; Blumenschine *et al.* 1994.
[23] Among others, see Domínguez-Rodrigo and Barba 2006, 2007; Blumenschine *et al.* 2007; also, Bunn 2001; O'Connell *et al.* 2002; Bickerton and Szathmáry 2011.
[24] See the various contributions in Vrba *et al.* eds. 1995; also, Kingdon 1993; Foley 1994, 1995; Potts 1996, 1998a,b, 2012; J. Allen *et al.* 1999; McManus *et al.* 1999; Taylor 1999; Alley 2000; Keeling and Whorf 2000; Stanley 2000; Richerson *et al.* 2001, 2005; deMenocal 2004, 2011.
[25] Potts 1996, 1998a,b, 2012; Potts and Faith 2015. (For more details on the causes and their consequences, see the outtake at my website.)
[26] See Kerr 1996; Hill 1987; also, Foley 1994; Potts 2012.
[27] White 1995, p. 378. White added: "It seems likely that the end of this time window was marked by technological innovation involved with the production of the Acheulean [tool] industry."

[28] De Menocal 2004, 2011. At the risk of offending the professional paleo-anthropologists, I have taken the liberty of lumping together the current (fluid) list of about a dozen early hominin species and treating them as categorically distinct from the dozen or so species of Homo, ranging from *Homo rudolfensis* to *Homo neanderthalensis* for the sake of clarity in explaining the major functional/adaptive changes over time for this audience. Some of the differences between "early" and later Homo species will be discussed later on.

[29] *Homo erectus* is the traditional name for the entire genus, or grade. Recent African fossil finds that are significantly different from Eurasian fossils have inspired the use of *Homo ergaster* for these African ancestors. Here, for convenience sake, I will use the original generic term for both. For a brief description of *Ardipithecus ramidus* see White *et al.* 2009.

[30] See especially Anderson 1986; Cheney and Wrangham 1987; Dunbar 1988; Cowlishaw 1994; Iwamoto *et al.* 1996; Wrangham and Peterson 1996.

[31] The evidence of predation against early hominins is discussed by Brain 1981, 1985. (For more on this, see the outtake at my website.) See also the Wikipedia article https://en.wikipedia.org/wiki/Dinofelis (last modified 9 June 2015).

[32] This assumption is derived from our common ancestry with chimpanzees, another primate that follows a pattern of male-based, multi-male/multi female groups. (For a critique of this scenario and my response, see the outtake at my website.)

[33] Wolpoff 1999a, pp. 217-219.

[34] First proposed by Morton 1927. See also Kurtén 1984; Coppens and Senut 1991; Wolpoff 1999a. Köhler and Moya-Solá (1997) report a recent find of a nine million-year-old fossil specimen, which they suggest may have used bipedalism for ground foraging in a more sheltered Sardinian environment. For a different point of view, see Isbell and Young 1996.

[35] Thorpe *et al.* 2007.

[36] Kingdon 2003.

[37] Kingdon (2003) believes that intensive terrestrial squat foraging would likely have been highly rewarding, judging by today's similar East African forest floors, which contain a rich variety of potential food sources. Squat feeding would have encouraged the necessary anatomical changes for bipedalism and would have freed up the hands to become effective foraging tools.

[38] Wrangham 1987.

[39] Diamond 1992.

[40] See Richmond *et al.* 2001; also, Kivell and Schmitt 2009.

[41] Sockol *et al.* 2007.

[42] Zollikofer *et al.* 2005. Indeed, the discovery some years ago of several 5.2-5.8 million-year-old *Ardipithecus* fossils in the Middle Awash area of Ethiopia indicates that the shift toward *Australopithecus* began far earlier than previously supposed. Some

features, from their bipedal gait to their distinctive teeth, clearly presage *Australopithecus afarensis* some two million years later. See also Lieberman 2013.

[43] For our purpose, the assortment of very early hominin fossils that have been given different names (*Ardipithecus, Australopithecus, Paranthropus*, etc.) can be lumped together under the general heading of australopithecine "grade". This will allow us to sidestep the controversies over labels, categories, and sequences and concentrate on the most general and important features.

[44] Silk 2011.

[45] Many other theorists over the years have also endorsed the group-defense model, including George Schaller, Alexander Kortlandt, John Pfeiffer, Carel van Schaik, Richard Alexander, Richard Wrangham, Joseph Henrich, and others. For instance, van Schaik (1983), in an in-depth study of non-human diurnal primates, concluded that security was the most compelling explanation for group living.

[46] These "smart monkeys," as baboon expert Shirley Strum (1987) calls them, are ubiquitous throughout Africa and the Near-East. See also Kummer 1968.

[47] See Alemsegad *et al.* 2006.

[48] For an analysis of the role of public goods in collective action across 135 animal species, see Willems *et al.* 2013. For an analysis of the role of cultural variation in group selection and human evolution, see Bell *et al.* 2009.

[49] For further analysis of this issue, see Cowlishaw 1994.

[50] The manifold advantages of size in evolution have been explored in depth by John Tyler Bonner 1999, 2003. See also the article by Adrian Bejan, "Why the bigger live longer and travel farther: animals, vehicles, rivers and the winds." http://www.nature.com/articles/srep00594

[51] See Gintis *et al.* 2015; also, Corning 2017.

[52] On this issue, see Wrangham *et al.* 1999; Wrangham 2009; Stanford 1999, p. 100.

[53] Gintis *et al.* 2015.

[54] Harmand *et al.* 2015; also, McPherron *et al.* 2010. See also http://news.sciencemag.org/africa/2015/04/world-s-oldest-stone-tools-discovered-kenya (accessed 23 June 2015).

[55] See especially the discussions in Lewin 1993; Kingdon 1993; Ambrose 2001.

[56] See especially Klein 1999; Wrangham *et al.* 1999; Wrangham 2009.

[57] See Shumaker *et al.* 2011.

[58] Pruetz and Bertolani 2007.

[59] Gintis and his colleagues agree. They believe that the use of stones as weapons may have been underrated; see Gintis *et al.* 2015.

[60] Does this mean that the australopithecines were "killer apes" after all? I would argue that this image is greatly overdrawn. (For more on this, see the outtake at my website, and the further discussion in Chapters 8 and 9.)

[61] Rose and Marshall 1996.

[62] See Wrangham 1987; Foley 1995; Stanford 1999.

[63] Among the recent writings on intergroup aggression among primates and evolving hominins, see especially Manson and Wrangham 1991; Wrangham and Peterson 1996; Stanford 1999; Corning 2007a; Gintis *et al.* 2015.

[64] See Wolpoff 1999a.

[65] Parish 1996; also, Tokuyama and Furuichi 2016. A detailed analysis can also be found in Wolpoff 1999a, pp. 141-142, 271-273.

[66] Discussed in depth in Sober and Wilson 1998.

[67] Holloway 1975, 1983a, 1996, 1997; Tobias 1971; Tobias, ed. 1985; see also Falk 1998; Falk *et al.* 2000; Ambrose 2001.

[68] See especially Conroy *et al.* 1998; Donald 1991; Aiello and Dunbar 1993; Dunbar 1996, 1998, 2001; Rilling and Insel 1999; Dor and Jablonka 2000; Ambrose 2001; Tomasello 2001, 2005, 2008.

[69] Dunbar 1996, 1998, 2001, 2009; also, Dunbar and Shultz 2007; Shultz *et al.* 2011; Sewall 2015. (For more on this, see the outtake at my website.)

[70] Tomasello 2014; also, Tomasello 2008. Tomasello explains that a shared intentionality involves a "we" orientation, which affects our relationships to others and to our environment. He believes this provides the basis for our collaboration and our cultural institutions, as well as our communications and language.

[71] See Kingdon 1993.

[72] Teaford and Ungar 2000. See also Alba *et al.* 2001.

[73] Toth 1987a, b; Toth *et al.* 1993; Schick and Toth 1993; also, Ambrose 2001.

[74] See Steele 1999; Sterelny 2012; Morgan *et al.* 2015.

[75] Potts 1998a,b.

Chapter 8

The Self-Made Man II: From *erectus* to *Homo sapiens*

There has been a tendency over the years to characterize the final emergence of *Homo sapiens*, beginning perhaps 300,000 years ago, as a "revolution," but there was a much earlier transition in human evolution that was equally transformational. It occurred some 2 million years ago, with the rise of *Homo erectus* and, presumably, other close relatives.

To be sure, this transition required hundreds of thousands of years and began modestly in earlier, precursor *Homo* species. Equally important, it was the combined result of a synergistic set of behavioral and anatomical changes over time rather than some "megamutation" or a "punctuated equilibrium."[1] But, taken together, these changes were of far-reaching importance; they established the foundation for the final evolutionary breakthrough that followed.

The story begins with *Homo habilis,* along with (probably) *Homo rudolphensis,* the recently discovered *Homo naledi,* and perhaps others.[2] Over the course of roughly three quarters of a million years, a major transformation occurred that set our ancestors firmly on the path that has led to modern humankind.

Although there are important differences among these ancient hominins, they represent to varying degrees an interrelated set of changes (both behavioral and anatomical) that were intermediate between the diminutive, small-brained australopiths and the modern-day, large-brained, technologically-sophisticated, language-using *Homo sapiens.*

At the risk of offending the professional fossil-hunters, I will treat *Homo erectus* as a generic category here and focus on the broader "why" question. Why did the hugely successful australopiths evolve into something much more suggestive of modern humans?

Stone Tools

A benchmark that is sometimes used to denote the transition between australopithicines and the *Homo* line is the appearance in the fossil record of the first flaked stone tools that were struck from larger quartzite or lava stone "cores." (Other commonly-used criteria are differences in brain size, dentition, and the like.)

However, it was discovered some years ago that *Australopithecus garhi* (or an as yet unidentified contemporary) at Gona, in Ethiopia, was already adept at manufacturing and transporting these implements over some distance and using them for a variety of purposes – chopping, cutting, smashing bones and perhaps skinning – at least 2.5 million years ago.[3] More recently, even this date has been pushed back much further with the discovery of very similar tools in Kenya dating to 3.3 million years ago, coupled with a discovery nearby of animal bones with many cut marks – clear evidence of ancient butchery.[4]

It has been commonplace for writers about human evolution to belittle the so-called Oldowan tool-making tradition – so named because the first stone tools were discovered at the Olduvai Gorge by Louis Leakey (father of Richard) and his co-workers decades ago. The naysayers claim that the earliest stone tools were not very sophisticated, and that there is little evidence of further technological progress for another million years.

But this attitude is condescending. I have already mentioned the demonstration by Toth and Schick that the manufacture and use of these stone tools is a more highly skilled activity than it appears. More important, anthropologist Bernard Campbell has argued that, contrary to the belittlers, the Oldowan tools were so successful, and so well-suited to the needs of a highly mobile biped, that there was probably no need nor incentive for making any major improvements. Leakey himself demonstrated how efficiently a large animal can be skinned and de-fleshed with a simple stone flake. More elaborate tool-making is also a very time-consuming activity. The more sophisticated tools of the later Acheulean tradition (dating back 1.8 million years or perhaps a bit earlier) required some 25 steps, while the more highly diversified and stylized Aurignacian tools of the Upper Paleolithic (within the past 50,000 years) required six separate stages and about 245 steps.[5]

A Technological Revolution

The evolutionary importance of these Oldowan tools can hardly be overstated. It amounted to a technological revolution, because it enabled our ancestors to become systematic hunters (and scavengers) and to exploit the many herds of large animals that had come to populate the open grassland areas in East Africa. To stress a point made earlier, tools create powerful synergies that, in turn, may be favored by Synergistic Selection.

Once stone tools were deployed by these early hominins, moreover, the carcasses of the animals they killed provided raw materials – horns, bones, skin and sinew – for many other uses as well. Just as digging sticks and hand-held weapons may have played a key role in the success of the early australopithecines, the invention of stone tools vaulted our ancestors into a new ecological niche. Equally significant, this adaptive revolution seems to have predated the emergence of *Homo erectus* by several hundred thousand years. In other words, synergistic behavioral changes – and technological innovations – preceded and supported the major anatomical developments that are reflected in the fossil record much later on.

The specimens of early *Homo erectus* that began to appear about two million years ago, were strikingly "improved" anatomically. First and foremost, they were about twice the size of Lucy and her cousins (and even greater than that by weight), or approximately equal in height to some modern *Homo sapiens*. Their brains were also double the size of the early australopithecines (about 800 cubic centimeters, growing later on to as much as 1100 cc's), and they show evidence of further reorganization and greater brain lateralization, or asymmetry, which is associated with handedness and other cognitive specializations.

In short, the hominin social brain had become both larger and more complex.[6] *Homo erectus* also had longer legs (and shorter arms) and had perfected the striding gait of modern humans. Not only was this much more energy-efficient than the knuckle-walking technique of chimpanzees (as noted earlier), but it was especially well suited for longer distance travel and even running. Whereas chimpanzees in open areas may have ranges of 200 square miles, human hunter-gatherers typically exploit ranges of close to 700 square miles.[7]

The Nuclear Family?

Another striking change in *Homo erectus* was a decline in the size difference between the two sexes to approximately modern human proportions (15–25% compared to about 50% among the australopiths). To some theorists, this suggests that sexual competition among the males had declined. Perhaps the nuclear family and more permanent male-female pair-bonding had emerged – just as Lovejoy proposed, only much later and for other reasons.

But it is also likely that larger females had significant functional advantages. One advantage would have been the ability to travel at a faster pace and not hold back the rest of the group. Another would have been the ability of the females (and the group) to cover more ground and greatly expand their day ranges. Still another advantage might have been the ability to carry heavier infants over longer distances. Larger size also implies a larger birth canal, an important consideration in giving birth to larger infants. Not least, it would also have been helpful for self-defense whenever the females were foraging independently, or if they were sequestered at home bases (see below).

A final distinguishing feature in *Homo erectus* – at least in the more delicately proportioned "gracile" forms – was a significant reduction and refinement of the teeth and mandibles and changes in wear patterns, along with a major reduction in the size of the digestive tract. These changes suggest yet another behavioral pacemaker – a dietary shift to one that required less chewing of coarse plant foods, much greater consumption of meat, and possibly cooked plant foods as well. In turn, these anatomical changes facilitated the later development of the vocal tract and the evolution of language skills (more on language below).

An Economic Revolution

What is the meaning of all this? The most plausible explanation, many theorists believe, is that a major behavioral shift had taken place – a Lamarckian change of "habits" that amounted to an economic revolution. This shift involved *four key innovations* that were synergistic and mutually supportive (a synergy of synergies), I would suggest.

First, in the half million years after stone tools became a standard part of their tool-kit, our hominin ancestors made the transition from a small forager that only opportunistically hunted and scavenged meat to a systematic (cooperative) hunter and confrontational/power scavenger that relied on meat (along with an array of plant foods) to provide a more abundant, high-quality, cost-effective food supply. *Homo erectus* had joined the ranks of top carnivores and could hold their own in confrontations with other carnivore competitors – not to mention potential predators. This conclusion is not original, of course, and there are alternative theories. But it seems most consistent with the anatomical (and technological) changes that occurred.[8] Needless to say, this development was highly synergistic.

A catalyst for this change may have been a significant shift in global climate to a more variable, oscillating pattern between 2.8 and 2.4 million years ago, and again after 1.8 million years ago.[9] One consequence for various East African inhabitants was a more arid, more seasonal environment with less abundant (more widely scattered) plant foods and many more herds of large game animals. A shift to the systematic exploitation of meat on the hoof was the most cost-effective response to these ecological changes. Organized cooperative hunting did not require a major behavioral change. These hominin entrepreneurs were already terrestrial, group-living, opportunistic hunters, and they certainly had other pack-hunting models for inspiration.[10]

Nor was big game hunting necessarily a dangerous activity, as many theorists have supposed. There are numerous less hazardous techniques. As Kingdon notes, coordinated tactics of various kinds can be used to ambush prey, or to panic and drive animals into mudholes, swamps, cul-de-sacs, deadfalls or even (in later times) into prepared trip lines, nets, and traps. Later on, fire brands were probably also used for driving potential prey, as well as for deterring carnivore competitors. Spears and clubs surely were also employed.[11] In other words, a synergy between intelligent problem-solving, close social cooperation, and effective use of tools/weapons most likely played a major role in the shift to a full-time hunting/scavenging way of life.[12]

A Nutritional Revolution?

Indirect evidence for this synergistic behavioral shift can be found in the very anatomical changes that occurred in *Homo erectus*. Their larger size and bigger brains imply that they had much greater nutritional requirements. Anthropologists William Leonard and Maile Robertson have estimated that, compared to the australopiths, *H. erectus* needed 40–45% more energy to support their increased size, and that larger day ranges would have further increased their total energy needs by as much as 85% (judging by contemporary human foragers).[13]

Females were especially burdened by the increased nutritional demands of their larger but quite helpless neonates, while a lengthening period of childhood dependency imposed greater constraints on the females' mobility and foraging ability. Anthropologists Leslie Aiello and Catherine Key have calculated that *Homo erectus* females may have needed some 60% more calories for reproduction than did their australopithecine predecessors. Aiello and Key point out that this would have necessitated a nutritional revolution.[14]

There are only so many hours in each day that can be allocated to food acquisition, so the anatomical changes in *H. erectus* put a premium on obtaining large quantities of high-quality food as efficiently (and dependably) as possible. To put it in economic terms, an increase in demand required a corresponding increase in supply. Systematic hunting/scavenging and provisioning of the group with meat was at once a major cause and a major part of the solution to this problem. Meat provides twice as many calories as fruit and ten times as many calories as leaves, not to mention proteins, fats, and other nutrients. Meat can also be obtained in large packages that allow for collective acquisition, bulk transport and shared consumption. Evolutionary biologist Daniel Lieberman and his colleagues, who favor a long-distance, endurance running model, have shown that this strategy would also have been highly cost-effective in terms of the net calories returned.[15]

Other scenarios are also possible, of course, but the hunting/scavenging scenario seems most consistent with other evidence – tooth wear patterns, tool use patterns, and the anatomical changes that occurred in *H. erectus/H. ergaster.*[16] Over the course of time there were also improvements in tool-making skills (as reflected in the Developed Oldowan and Acheulean traditions), plus more selective use of raw materials, more complete processing of animal carcasses, and evidence of more specialized tools for different uses, such as wood working, skinning, and plant food preparation.

It's also likely that there was an increased need for transporting food, stone tool cores and, most important, water supplies over long distances. In fact, water became a much more critical resource with the adoption of hunting and scavenging activities in an open savanna environment. As anthropologist Rosemary Newman has pointed out, we evolved into a "thirsty, sweaty animal" that needed as much as two quarts of water per hour while engaged in the hot pursuit of animal prey on a hot savanna day.[17]

Home Sweet Home

However, there is a second, related behavioral change that was also critically important, it would seem. Secure home bases or encampments, even if temporary, became a functional imperative under the big game hunting/scavenging scenario. If the males shifted from a joint foraging strategy to a strategy based on much longer, free-ranging searches, along with the pursuit and killing of larger, fast-moving animals, the females and infants could no longer tag along and the males could no longer serve as full-time guardians for the group. The obvious solution was to sequester the females and children, and to provision them with meat from their kills.

The evolutionary significance of this development was that it resulted in a major improvement in economic efficiency for the group as a whole, because it allowed for a more elaborate combination of labor and a more efficient use of time in a protected environment. Resources as needed – meat, plant foods, stone tool cores, animal skins, water, firewood, etc., – could be carried to a safe haven and then shared and utilized through a network of reciprocities.

Various theorists have noted that the ability to carry things over a long distance is an underrated technological achievement, perhaps because it involves a soft technology that does not fossilize. Yet it was a key innovation. Most critical, perhaps, was the need to import water when the group was not camped close to a stream, or pond, or lake. (Many of these water sources probably had to be avoided as campsites because they were magnets for other large predators.)

Perhaps the first baby-sitting cooperatives (allomothering) also date to this era. Allomothering would have freed up some of the females for local foraging and, later on, for fire-tending and the daily quest to find firewood. Indeed, anthropologist Sarah Blaffer Hrdy has proposed that allomothering was an indispensable (synergistic) enabler that played a critical role in the evolution of *Homo erectus*.[18] She believes it was a prerequisite for meeting the increased nutritional demands associated with more frequent child spacing (compared to chimpanzees), as well as a longer childhood dependency and a greater need for cognitive development. In Hrdy's words: "Alloparental care and provisioning set the stage for children to grow up slowly and remain dependent on others for many years, paving the way for the evolution of anatomically modern people with even bigger brains… Bigger brains required care more than caring required big brains."[19]

Homo pyrotechnicus

A third major innovation, the acquisition of fire by our ancestors, tends to be underrated these days, perhaps because it is a veritable cliché about human evolution. Yet the adoption and controlled use of fire was a truly revolutionary change. Darwin himself considered it to be one of our greatest inventions.[20] Natural fires caused by lightning or the combustion of gas seepages, etc., had been a part of our ancestors' environment for many millions of years. Even the earliest hominins no doubt observed the effects produced by fire, especially as a means for driving away other animals and as an instrument for killing and cooking potential prey. It's likely that the australopithecines also learned to scavenge at fire sites, just as their carnivore competitors and predators were doing. In fact, fire may have begun to play a major part in our evolution much earlier than we thought.[21] The so-called Karari sites analyzed by anthropologist Randy Bellomo suggested that hearths were used, most likely by *Homo erectus*, at least 1.6 million years ago.[22] And Richard Wrangham and his colleagues go back even further. They believe there are indirect "signals" of controlled fires as early as two million years ago, although others are skeptical about such an early date.[23]

In any event, over the course of time fire came to have many valuable uses – defense against predators, chasing competitors away from carcasses, tenderizing meat, killing harmful bacteria, breaking down toxic chemicals in the many plant foods that cannot be eaten raw, hardening wooden tools, drying skins, deterring insects, providing warmth (especially in colder, temperate climates), and even facilitating long-distance signaling and communications. Later on, our ancestors also learned how to use fire to condition the environment – to clear land and stimulate the growth of favored species and to prepare garden plots for planting. Not only was the controlled use of fire an immensely useful, synergistic technology – one without precedent in evolution – but it became an indispensable adaptation. We are also, in a real sense, *Homo pyrotechnicus*.

Food Processing and Cooking

The fourth major innovation, one that (literally) helped fuel the evolution of *Homo erectus*, was the adoption of food processing and cooking – just as Richard Wrangham and his colleagues have proposed (see

Chapter 7).[24] The timing is debatable, of course, but the logic is compelling. The ever-growing demand for a larger food supply, plus the many obvious advantages of food processing and cooking – rendering both meat and tubers far more easily chewed and digested, among other things – combined to make processing and cooking highly rewarding "soft" technologies. (It also allowed for a greatly reduced digestive system, thus economizing on an "expensive tissue.")

Wrangham points out that a sustained raw food diet is incompatible with a high-energy life-style and that *H. erectus* would have had to spend all day chewing in order to meet its calorie needs from raw, unprocessed plant foods alone.[25] He also notes that processing and cooking actually increase the net energy value of plant and meat foods. If sequestered home bases with fires were already an established behavioral pattern, the adoption of cooking represented only an incremental further innovation (Occam's razor).[26] The burden of proof should be placed on those who say that cooking did not arise at a relatively early date.[27]

A Synergistic Package

These and many other soft technologies – nets, weirs, traps, spears, shelters, containers, rafts and more – most likely long predated the technological achievements of modern *Homo sapiens*. The significance of these synergistic inventions was that they enabled *Homo erectus* to generate food surpluses, which allowed for the growth of bigger, more mobile groups of physically larger, longer-lived individuals and, equally important, their more numerous offspring.

Indeed, the population growth curve in hominins may have started to rise as early as two million years ago. Evidence for this (though circumstantial) is the fact that, at a relatively early date, *H. erectus* began to migrate out of Africa and to colonize other tropical and subtropical areas at great distances from their East African birthplace. Hominin fossils tracing back to between 1.8 and 1.6 million years ago have been found in Indonesia, and some fossil remains of that vintage have also been found in the former Soviet republic of Georgia (north of modern Turkey and western Iran) and at Ubeidiya in Isreal. These very long-distance travels were no small achievement; every new environment presented novel challenges and opportunities.[28]

In sum, it was our remote *Homo erectus* ancestors who invented the fundamental survival strategy of our genus – close-knit cooperation, an egalitarian social structure, an expanding division/combination of labor, along with the invention of various supportive technologies and a shared cultural tool-kit. *Homo erectus* was the product of an interrelated package of synergies that laid a foundation for the many improvements that have followed. For starters, there was a synergistic relationship between an improved, more efficient bipedalism, increasingly skillful, manipulative hands and the increased mental powers needed to use them in more productive ways. There was also the social triangle – a framework of cooperative relationships among the males, between males and females, and among the females. This scaffolding allowed for the further development of the group as a unit of collective adaptation over time, with greater organization and a more complex division of labor.

Culture in *Homo erectus*

The invention of a more efficient social organization and new technologies was only half the story, however. Cognitive development and culture – the accumulated know-how and experience of the group – also became an indispensable part of the package that enabled *Homo erectus* to prosper and ultimately to colonize new environments. Psychologist Steven Pinker calls it the "cognitive niche," while anthropologist Joseph Henrich characterizes our species as "cultural learning machines." He stresses that cultures are themselves potent "tools" that collectively empower social groups.[29] Larger cooperating groups are also able to exploit many new opportunities for social synergy, including sharing costs and risks, pooling information, jointly solving problems, developing new technologies and new combinations of labor and, not least, benefitting from synergies of scale against competitors, predators, and prey. Likewise, mutual aid, or "succorant behaviors," can increase the odds of surviving an injury or illness, while the policing of bullies and "free riders" serves to reduce internal conflicts – as noted in Chapter 3.[30]

Some theorists have claimed that culture only arose much later on in human evolution, with the emergence of art, rituals, symbolic language and the like. But if culture is defined more broadly as a body of socially transmitted knowledge, skills, tools, and other artifacts that are passed down from generation to generation through learning and teaching, then culture was already a vital part of the hominin life-style even among the australopithecines.

The very existence of a tool-making "tradition," after all, implies the ability to transfer the requisite skills between generations. And stone tools were surely the tip of the iceberg. For instance, modern hunter-gatherers have "mental maps" of hundreds of square miles and can recall the precise locations for literally of thousands of water holes, plants, animals, natural hazards and landmarks, most of which they have learned from their elders.[31]

"Hominin Nature"

Undergirding and supporting this expanding cultural niche were significant changes in the "psychological anatomy" of our ancestors – a growing suite of pro-social motivations and emotions (not to mention various cognitive skills). Call it our "hominin nature." These emergent social traits included such things as a readiness for sharing with others, reciprocating for acts of generosity (coupled with strong antipathy toward free riders and cheats), the ability to empathize with the feelings of others and a sense of fairness in our dealings with them, a powerful inclination to follow leaders (sometimes slavishly and indiscriminately), a willingness to follow rules and norms of behavior, and a strong tendency to bond emotionally with our group.[32] We would call it "patriotism."

More than a century of research by child psychologists, from Jean Piaget to Lawrence Kohlberg and, currently, the comparative psychologist Michael Tomasello and his colleagues (among others), has confirmed that these and other social traits emerge spontaneously in human children, although they are obviously shaped by cultural influences as well.[33] As Tomasello, Henrich and others stress, human nature and human cultures have co-evolved. Tomasello highlights especially the role of "shared intentionality" – common objectives – in facilitating the evolution of cooperative problem solving and cooperative behaviors.[34] Henrich points to the rise of what he calls the "collective brain," our shared cultural capacities. He notes: "Our collective brains arise from [the] synergies created by the sharing of information among individuals." He also describes our cultural evolution as "a broad and synergistic process."[35]

Beginning with *Homo erectus*, culture became an increasingly potent adaptive tool. New ideas and inventions were not only preserved and communicated to subsequent generations but they were also improved upon over the course of time.[36] In effect, the group as a whole became a repository of adaptive information and an engine for the invention of more cultural synergies. Spears, for example, came to be made of better raw materials; they were more finely shaped and balanced; their tips were fire-hardened; barbed tips were added to increase their penetrating and holding power; wooden spear throwers (atlatls) were devised as a way to increase their range, striking force and accuracy; finally, bows and arrows were invented as a lightweight alternative that could increase a hunter's range and precision, and (not least) multiply his supply of "ammunition."

Each of these inventions represented a major economic advance. More food could be acquired more dependably with less time, effort, and collective risk. Chistophe Boesch and Tomasello call it a "ratchet effect,"[37] while Henrich speaks of an "autocatalytic" process of reciprocal causation (co-evolution) between cultural and genetic change. Henrich also reminds us that the process is not linear; technologies and other cultural attainments can also be lost.[38]

The Evolution of Language

The backbone of modern human culture is, of course, symbolic language – the ability to communicate complex information among members of a group (and beyond) both rapidly and in precise detail. Language gives us an enormous collective advantage that no other species enjoys. It is indispensable for a complex society – a classic example of necessity being the mother of invention, and it played a vital role in our evolution long before the full development of symbolic thought and articulate speech. There is good reason to believe that language (broadly defined) is fundamentally a cultural innovation that has been progressively improved over time with both anatomical changes and further cultural inventions – much like the progressive development of bipedalism, our skillful hands, our cognitive abilities, and our technologies. Indeed, language is a flexible tool that is still evolving even today.

The origin of language has been debated literally for centuries. Like the problem of explaining human evolution itself, language is another minefield where the theories have far outrun the evidence. These days there are so many different explanations – some very reasonable and others quite far-fetched – that it is difficult to sort them all out. On the one hand, there are the so-called "discontinuity theorists" like the linguist Noam Chomsky, who claims that some "chance mutation" about 100,000 years ago, suddenly made language possible.[39] Others, like evolutionary psychologist Steven Pinker, see language as having evolved more gradually, although they also see it as mostly innate.[40]

An Essential Enabler

I am in a different camp, one that views modern symbolic language as an outgrowth of a more ancient and essential enabler for the close social cooperation that was pioneered by our earliest hominin ancestors. In fact, all social systems require cybernetic control and communications processes; information exchange plays an important facilitating role.[41] (Some theorists stress that language may also have other social functions, such as providing a social "glue" for a group, or creating social boundaries between groups.)[42]

The so-called functional approach to language has been hotly debated over the years, but some of the heat was the result of conflicting definitions about what constitutes language and its key properties. For instance, some theorists apply the term only to vocal communications that employ arbitrary combinations of symbols (morphemes). Others believe that advanced thinking is essential for true language to arise. I side with those who use the term "language" in a broad sense as a behavioral means for communicating information; it is not about thinking, or symbols per se, but about intentional "signs" – to use the organizing concept of the science of semiotics.[43]

From a functional perspective, many other species, especially our primate cousins, have at least rudimentary "language" skills. Vervet monkeys, for instance, utilize an array of calls to differentiate among various categories of objects, from predators to fellow vervets. Naked mole rats use an elaborate system of vocalizations to organize their complex social life.[44] Our closest relatives, the chimpanzees, are especially impressive communicators. Their social interactions are shaped by a sophisticated array of vocalizations, facial expressions, gestures and body language. Perhaps most remarkable is the recent report that the Japanese great tit (a passerine bird) can create "compositional" calls (containing more than one element) that utilize grammatical rules, or syntax.[45] Recall also the multifaceted communication system in leaf cutter ants described in Chapter 4.

Language as a Necessity?

Thus, it seems unlikely that australopithecines and (even more certainly) *Homo erectus* could have developed a highly-organized pattern of cooperation and group adaptation – including a division/combination of labor, increased mobility and, ultimately, long distance migrations – without communications skills that, at the very least, greatly exceeded the repertoire of chimpanzees (some 30 calls, along with a repertoire of gestures and body language). Like other unique human traits, the functional advantages even of rudimentary communications skills are likely to have preceded and induced various anatomical changes, including the development of our vocal tract.[46]

I'm partial to Steven Mithen's so-called "Hmmmm" theory. Mithen envisions that our remote ancestors combined vocal utterances with gestural, mimetic and even musical elements to communicate with one another. Mithen calls it a "holistic" system.[47] Indeed, such multi-modal communications systems are ubiquitous in other social species as well. Psychologist Charles T. Snowdon found that 11 of the 18 language criteria developed many years ago by the linguist Charles Hockett are utilized by other animals, and that the remaining seven can also be seen (though not together) in various primates.[48]

Another theory that has recently gained some traction holds that human evolution was shaped in part by a process of "self-domestication" – a behavioral selection dynamic that is similar to the way domesticated cats and dogs may have evolved.[49] This idea has also been applied to the evolution of human language, where it is proposed that a reciprocal feedback process was involved. Tomasello speaks of it as being a "dialectical process."[50]

A Lamarckian Process

In any case, human language development was very much a Lamarckian process (properly understood). Even limited verbal communications skills would have provided important functional advantages for hominin groups. And these synergies were the very cause of the natural selection over time of improvements in the anatomical features that supported more advanced language skills.

To thoroughly mix two well-known metaphors, Richard Dawkins' "blind watchmaker" (the governing metaphor in his popular book about natural selection) could not have produced Jared Diamond's "swiss watch" (his way of characterizing our vocal organ) without the guiding hand of purposeful behavioral changes over time by our resourceful, inventive, loquacious ancestors. Indeed, the immense anatomical complexity of our language system – with literally dozens of tiny muscles, bones, nerves, soft tissue and, no doubt, many millions of interacting neurons – make it extremely unlikely that it could be of recent vintage. In a very real sense, language and the anatomical basis for language co-evolved. (It's also increasingly evident that our cognitive and linguistic skills overlap, as various theorists have stressed.)[51]

Some "hard" evidence for this Lamarckian scenario can be found in *Homo erectus/Homo ergaster* fossils. Jeffrey Laitman and his colleagues see anatomical evidence for changes in the vocal tracts of these early hominins that would have facilitated a more elaborate form of sound production.[52] In the same vein, brain endocast researchers Ralph Holloway and Dean Falk interpret changes in the neuroanatomy of *Homo erectus* as suggestive of having greater language skills. Holloway, Falk and others see further improvements in this direction throughout the Pleistocene Epoch (most of the last 2.6 million years).[53] Likewise, the appearance of the language-related (and now famous) FOXP2 gene and the control of breathing (a necessary language prerequisite) have been dated to somewhere between 1.8 and .5 million years ago.[54] These changes do not, of course, tell us much about the causal relationships, but they do support the general argument that progressively improving language skills were an integral part of the *H. erectus* narrative.[55]

The case for this co-evolutionary view of human language is also buttressed by some indirect evidence. It is significant that the syntax (the relationships between morphemes, or words) and the lexicon (the specific collection of morphemes and their meanings) differ widely among different languages. These important "surface" elements are culturally determined. Moreover, each new generation of children must learn a distinct repertoire of sounds, as well as the syntax and lexicon for the particular language they are exposed to. It is especially significant that a child can learn any language, yet a severe language deficit will occur if a child is isolated altogether from a linguistic cultural environment. In short, language is inextricably both biological and cultural in nature.[56]

Zoon politikon

There is one other important behavioral development among our remote ancestors that is not so widely appreciated, I believe. Aristotle long ago dubbed humankind "*zoon politikon*" – the political animal – and political theorists ever since have used this evocative term as a touchstone.[57] What Aristotle was referring to is the fact that humans, like a number of other species, engage in organized cooperative activities that entail collective decision making, coordinated behaviors, and leadership.[58]

The traditional view among the students of animal behavior has been that the political systems in social mammals – say baboons – normally involve an authoritarian dominance hierarchy maintained by physical threats and coercion.[59] However, more recent research and theoretical work on the subject paints a much more complex picture. Leadership is at once more common than we thought in other socially organized mammals and is more varied and complex, with numerous examples of a more consensual, cooperative pattern.[60]

However, our own ancestors developed a system that is unique. In a series of landmark studies many years ago, the anthropologist Christopher Boehm documented the fact that modern hunter-gatherer societies are not only egalitarian but also exhibit what Boehm called a "reverse dominance hierarchy" – a social coalition that actively contains and suppresses aggressive individuals.[61] It appears that a more democratic political system may be a deep-rooted trait in our species.

More recently, the evolutionary economist Herbert Gintis, together with Boehm and the anthropologist Carel van Schaik, have proposed that this distinctively human behavioral pattern first evolved among *Homo erectus*.[62] They argue that the emergence of a more consensual leadership style in our ancestors was related to the adoption of cooperative big game hunting and confrontational scavenging with the use of "lethal weapons," perhaps as far back as 2 million years ago. Gintis and his colleagues believe that these weapons also provided important political "equalizers" for *H. erectus* that facilitated a more egalitarian social structure.

Australopithecine Politics?

Elsewhere I have argued that the shift to a more egalitarian and consensual leadership pattern – one that is based on "prestige" rather than dominance – might in fact be traced back even to the australopithecines.[63] The intense social interdependence required for a cooperative foraging strategy and the evidence of tool/weapon use and meat consumption as early as 3.3 million years ago, suggests that *zoon politikon* first emerged among these very early ancestors. The incessant daily demand for group decisions and actions related to foraging, nesting sites, water holes, opportunistic hunting and scavenging, dealing with threats, and, not least, migrations to new locations, all involved political processes. Competent leadership was vitally important, and there is much evidence that cohesive groups often make better collective decisions than individuals alone. The synergies derived from good decisions and effective social action may even have served as a potent selecting agency among different hominin groups (Synergistic Selection) – as Gintis, van Schaik, and Boehm also note in their paper, following Darwin's lead.[64]

It seems likely that the most powerful check against dysfunctional dominance behaviors in australopithecines was the potential for collective resistance by the group – the power of superior numbers (another synergy of scale) – in the same way that bonobo females gang up to thwart aggressive males. It may well be that the development of big game hunting and scavenging with advanced weapons in *Homo erectus* led to a further elaboration of our unique political system, but it seems likely that consensual leadership was already a well-established social pattern among our more remote ancestors. (The question of how this system was ultimately subverted in modern human societies, and the ominous consequences of this political regression will be addressed in Chapters 9 and 10.)

The Rise of Humankind

The image of modern *Homo sapiens* as the product of a giant leap of some kind – a "revolution" – seems irresistible. Christopher Wills, for example, characterized our species as "evolutionary speed-demons." He described humans as having a "runaway brain," which he likened to the fictional character Cyrano de Bergerac's outsized nose. Other theorists refer to a "creative explosion" (John Pfeiffer), a cultural "revolution" (Ian Tattersall), a "great leap forward" (Jared Diamond), a "punctuational event" (Richard Klein), and the like.[65]

However, the accumulating evidence seems to contradict this model. Rather than characterizing it as a sharp discontinuity, or a new stage, it would be more accurate to say that a synergistic threshold was crossed in an already quickening pattern of progressive functional improvements. Even the categorical distinctions between different tool industries over time are not so clear cut (pun intended) or static as was once thought to be the case.

But more important, during this long transitional period in human evolution – several hundred thousand years or more – there was a sequence of incremental biological and cultural changes that accelerated over time. Moreover, there was no single epicenter for these developments; it was a much more dispersed, mosaic pattern. In general, evolving "archaic" humans became more gracile, their teeth and jaws receded as their facial protrusion declined and, of course, their brains got bigger in relation to their body size (encephalization). The increase in brain size is especially significant, because it provides indirect evidence of further improvements in the ability to procure high-quality food resources.

A Cultural Decoupling

Perhaps most significant is the fact that the decoupling of biological and cultural evolution became more pronounced. For instance, the later Acheulean tool industry – including a more highly diversified array of "biface tools" shaped to consistent patterns – emerged several hundred thousand years before a major surge occurred in brain size, beginning perhaps 500,000 years ago. Likewise, the rise of the Mousterian tool industry some 250,000 years ago, including especially the development of composite tools such as mounted (or "hafted") wood/stone axes, may have preceded the emergence of fully modern *Homo sapiens* by 50,000-100,000 years. Large scale hunting with skillfully made throwing spears also became routine long before modern humans appeared.[66]

In other words, the development of new technologies, especially for food getting, may once again have preceded and supported the biological evolution of energetically more costly (and brainy) humans. Especially noteworthy is the fact that the anatomy of the "late model" *Homo erectus* was very close to that of *Homo sapiens*. Among other things, the brain size of *H. erectus* in the late Pleistocene was well within the range of modern humans.

In short, the image of a "revolution" is a serious overstatement.[67] Both biologically and culturally, evolving hominins made many functional improvements throughout the Middle Pleistocene (from about 750,000 to 250,000 years ago) and beyond. The momentum toward modern, culturally complex humans was already well-established and was accelerating when our most immediate ancestors emerged, perhaps 300,000 years ago. As paleoanthropologist Milford Wolpoff notes, "modernization... was an ongoing process in widespread populations and not a single dramatic event in a single population."[68] His colleague Richard Klein agrees: "Middle Paleolithic people were advanced over their predecessors in many ways."[69]

Alternative Paths to Modern Humans

Of course, this begs the question. What caused the dramatic final sprint that vaulted our ancestors into a unique new evolutionary niche – the most recent major transition in evolution? And what happened to *H. erectus* and the other archaic *H. sapiens* species, including the Neanderthals and the Denisovans? More intense even than the debate over the evolution of language is the ongoing controversy over the crossing of the final Rubicon to modern humankind. As the paleo-anthropologist and human evolution expert, Chris Stringer, concedes: "We have partial answers to the 'when' and 'where' of our origins, but still precious little about the 'whys.'"[70]

One theory, the so-called multi-regional model, is most closely associated with Milford Wolpoff and his colleagues, although other versions of it have also been proposed.[71] In Wolpoff's view, the array of evolutionary trends that were evident throughout the Pleistocene continued uninterrupted into the late Pleistocene/Upper Paleolithic without any sharp transition. The essential unity of the process, he believes, was assured through a pattern of both gene flow and cultural exchanges between regional populations. Wolpoff illustrates the idea with the image of various stones being thrown into a pond; as the ripples from each stone spread outward, they interact with each other.

Although this theory is controversial, it has also been widely misunderstood. Wolpoff and his co-workers did not propose that there was an independent invention of *H. sapiens* in different places, and that fully modern humans arose more than once. Wolpoff himself pointed this out many years ago.[72] Rather, their claim is that humankind did not arise in Africa but emerged from a broadly diversified geographic area, and that the regional differences found in humans today represent an amalgamation of a deep heritage rather than a more recent divergence from a single "invading" stock.

Recent DNA testing suggests that there was in fact some interbreeding (or hybridization) between modern humans and their antecedents and close relatives in various places.[73] All of us carry some 2 to 5 percent of the Neanderthal's genome. In addition, many artifacts have been unearthed in recent years that narrow (if not eliminate) the cultural distance between *Homo sapiens,* the Neanderthals, and other archaic humans.

Out of Africa?

Nevertheless, the currently favored model for the final emergence of modern *Homo sapiens* is the "single origin" or "out-of-Africa" hypothesis, long championed by Stringer and others. An increasingly compelling body of genetic evidence – mitochondrial DNA and Y chromosome data in particular – indicate that all modern humans trace their lineages back to a common ancestor (perhaps a very small "bottleneck" population) in Africa about 100,000 years ago, or perhaps even more recently.[74] The data indicate that this founding population grew larger over time and began to migrate out of Africa, beginning about 60,000 years ago (or maybe somewhat earlier).[75]

In other words, various genetic "markers" reveal that there was after all an epicenter – a starting line for the final lap to humanity – and that modern humans effectively replaced other hominins in various parts of the world in short order, including the Neanderthals in Europe and the Middle East, the Denisovans in Europe and Southeast Asia, and perhaps others as well. It seems there was an exodus, a wave-like advance by an expanding population that spread their unique genes around the world and, over time, displaced the earlier colonizers. How can we account for this human tidal wave?

Our understanding of this momentous transition starts with the growing evidence that our final emergence was not the result of a linear development but was preceded by a messy, geographically dispersed pattern of temporary advances, reversals, and reappearances, as our ancestors were forced to adapt over time to a challenging and changing environment. Chris Stringer admits that "I am reconsidering many of my previous views on the origin of our species in Africa, and now I think we need to talk about *origins*, rather than a single point of origin."[76]

It's also now evident that the outcome was shaped by a synergistic convergence of several factors – a complex interplay of ecology, population size, technological innovations and other cultural advances. Recent evidence suggests that *Homo sapiens* may have become widely dispersed throughout Africa long before the great migration out of Africa began.[77]

Some students of human evolution over the years have elevated language acquisition to the status of being a prime mover – or at least *primus inter pares* – in the final emergence of *Homo sapiens*. For instance, cognitive scientist Philip Lieberman has argued that symbolic language, or rather the lack of it, played a decisive role in the final demise of the Neanderthals. Lieberman did not explain exactly why.[78] In the same vein, Jared Diamond asked rhetorically, what was the "missing ingredient" in human evolution – the key that can account for what he characterized as the "great leap forward" during the past 100,000 years? The answer, he said, was "the perfection of spoken language."[79] Ian Tattersall, likewise, called language "the fount of our creativity."[80] And Richard Klein speaks of a "genetic revolution."[81] This was the innovation, presumably, that opened the floodgate of human inventiveness and liberated our big brains from their cranial cages.

A More Complex Story

Such monolithic explanations are no longer tenable, I believe. The emerging story is much more complex. One major factor was the influence of severe, disruptive climate changes during this period. The last major Ice Age began about 125,000 years ago, intensified about 33,000 years ago, and peaked about 20,000 years ago (as noted earlier). Within this broad global trend, moreover, there were a number of spikes in local and regional climates, most notably one that occurred at about the same time and place where the exodus from Africa is presumed to have begun.[82]

However, the critical difference, once again, had to do with how various emerging *Homo sapiens* groups responded. Some of them atrophied or disappeared while others adapted and were ultimately able to thrive and increase their numbers. The well-known anthropologist-ecologist team, Robert Boyd and Peter Richerson, along with some of their colleagues, have advanced a cultural explanation. They argue that our ancestors' increasingly sophisticated cultural capacities – not genetics – played a decisive role in the final sprint to humankind. Some human groups acquired an unprecedented ability to coordinate their actions, invent new solutions, and, most important, adapt rapidly and flexibly to various challenging circumstances.[83] Long-distance migration of human groups into radically new environments, often under severe ecological stress, represented an extraordinary accomplishment.

An Economic Explanation

I would emphasize the economic (and synergistic) basis of this transition. Why were *Homo sapiens* able to expand their numbers while other contemporary hominin groups were static or declining? The difference had to do with the development of a superior economic system, one that could produce the surpluses needed to feed a growing population. It is likely that the migrants from Africa possessed a synergistic package of cultural traits that tipped the "balance of power" with competitors who otherwise were very similar. This package included a number of technological and social/cultural innovations, along with a major demographic advantage (larger population sizes). It's possible that they were aided by improved language skills, although language was certainly not the whole story.

It's not coincidental – many theorists believe – that the timing of the African exodus coincided with the flowering and spread of a cluster of technological and social innovations.[84] These included more diversified and specialized tools made from various materials, more skillful manufacturing techniques, better cooking skills, more elaborate shelters, better food storage capabilities, greater use of marine resources, longer occupation of different camp sites, greater mobility, and, especially important, larger numbers (approximating the population densities of modern hunter-gatherer societies).[85]

Richerson, Henrich and others argue that group size determined the degree of cultural complexity, but the reverse was equally true; advances in the economic system undergirded and ultimately determined group size.[86] In other words, the human "revolution" was a Lamarckian process; the rapid growth and world-wide spread of humankind was at bottom an economically-driven process, utilizing new "habits" that exploited new synergies.

Just to illustrate, a study undertaken by economic anthropologist Ralph Salisbury in the 1960s documented that a New Guinea native equipped with a steel axe could fell five times as many trees in a given amount time as his ancestors could with stone axes. Likewise, in a study that compared two native populations in Venezuela, it was found that a shot gun was two to three times as efficient as a bow and arrow at killing animal prey.[87] In the Paleolithic, the prime candidates for major technological game-changers were (1) spear throwers, which greatly increased the range and accuracy of the users; (2) more powerful composite wood/stone axes; and (3) quite possibly, bows and arrows.[88]

Pre-historic Warfare?

Not only did a more potent cultural package provide an important economic advantage but it gave our African ancestors a major "military" advantage as well. It seems likely that the human diaspora of 50–60,000 years ago, was at times a hostile incursion into already occupied lands rather than a peaceful trek into virgin territory. The human wave was often (though perhaps not always) accompanied by coercion and even violence. This is not a new theory (as noted in Chapter 7), but it deserves a new look.[89]

I hasten to add that we are not talking about wars of conquest or imperialism in the modern sense. The terminal Pleistocene humans were not necessarily more warlike in temperament, nor were they seeking dominion for its own sake. It's more likely the process was driven by a pressing need for resources to support a growing population in a highly changeable environment. Call it the resource acquisition model of human aggression. Nor would it have included set-piece battles resembling modern warfare. It's more likely to have involved a suite of coercive measures – intense competition for food resources, pointedly threatening displays of force, ambushes, raids and the like – that were used to intimidate and/or displace competitors and, probably, to acquire females. (Like their remote ancestors, emerging *Homo sapiens* seem to have been organized as male-based groups).

It should go without saying that a conflict scenario in any form is a distasteful idea, but there are a number of reasons for suspecting that coercion played a major role in the prehistoric dispersion of humankind. One is the evidence, based on a large body of research, that ethnocentrism (loyalty to one's own group) and xenophobia (hostility between groups) is an evolved psychological propensity in humankind with a strong biological foundation and deep evolutionary roots.[90] In fact, xenophobia is also a common trait in social primates, especially the chimpanzees. Many examples of raiding and conflict between groups have been documented among our closest primate relatives, not to mention the extensive evidence of collective violence among surviving human hunter-gatherers.[91] There is even some fossil evidence of warfare in pre-historic times.[92]

Another reason for suspecting that coercion was involved in the final spread of humankind is that expanding human populations were not migrating into uninhabited territories, for the most part. The humans that trekked out of Africa discovered that other hominins had preceded them by hundreds of thousands of years. Peaceful trade relationships can arise when there is the possibility of mutually beneficial exchanges between groups. But any attempt to seize another group's territory and resources is a zero-sum game. The immigrants came as competitors who eventually threatened the livelihood of the established residents. (As noted above, there was evidently some gene flow, or hybridization between these populations as well.[93] Some theorists portray it as a love-in, but there's an old saying among geneticists that conquerors have always been very generous with their genes.)

A Synergy of Scale

Equally significant, the evidence suggests that migrating human groups were able to mobilize and coordinate larger numbers than their competitors – a synergy of scale that has often served as an incentive for collective violence and very often determines the outcome.[94] A disturbing example, cited by Jared Diamond in his Pulitzer Prize-winning book *Germs, Guns and Steel*, involved the total destruction of the small Moriori hunter-gatherer society on the Chatham Islands (in the Pacific) in 1835 at the hands of 900 well-armed Maori agriculturalists from nearby New Zealand. The Maori first learned of the peaceful Moriori from a transcient Australian seal hunter. Aroused by the report that the Moriori had no weapons, the Maori immediately organized a seaborne invasion. When the unsuspecting Moriori didn't resist, the Maori raiding party slaughtered them with impunity.[95]

A further military advantage for the migrants out of Africa, most likely, was that they enjoyed a superior military technology – including any one or all of the "candidate" innovations mentioned above: spear throwers (atlatls), advanced stone axes, and/or bows and arrows. The value of these weapons in personal combat against a simple wooden club, or a hand axe, or a crude spear is obvious. It is noteworthy that the first (known) evidence for the use of these more advanced tools roughly coincides with our best estimate for when the human migration out of Africa began.[96]

This scenario is also consistent with the pattern in more recent history, where human conquests have typically involved some combination of larger numbers, better organization and/or a decisive technological edge. The implication seems clear. The human diaspora most likely included many episodes of coercion – the forcible containment and displacement of the prior occupants by modern humans. Given the inherent risks, however, it's likely that violent confrontations were more episodic than endemic. Indeed, sometimes migrant *Homo sapiens* and the resident hominin groups appear to have co-existed for a time (as was clearly the case with the Neanderthals), but eventually the earlier arrivals disappeared. The biological term for this is "competitive exclusion." (A suggestive recent study in Southern France found that, at about the time when the Neanderthals vanished from Europe, there were ten times as many human settlements as Neanderthal sites.)[97]

A Convergent Scenario

The well-known anthropologist Curtis Marean has recently proposed a theory to explain the human diaspora that is convergent with the scenario described above but also differs somewhat.[98] He believes that there was an expansionist (and militaristic) "invasion" of Europe and Asia fueled by a "new" combination of "hyperprosociality" – intense cooperation among non-relatives – and the "breakthrough" technology of spear throwers. In contrast, I see close cooperation (including with non-kin) as an already long-established social pattern and give more weight to economic advances and the ability to support larger groups with more efficient food getting technologies and techniques. Marean points to the role of "dense and predictable resources," but I would emphasize the cultural innovations that were necessary to identify and exploit these resources.[99] Superiority in numbers, to repeat, also provided an important military advantage – a synergy of scale. Thus, it was the ongoing need to support larger groups, and the relentless daily food quest in a changing environment over time that constituted the major drivers for the human migrations. It was also, needless to say, a process that spanned many thousands of years. In short, demographic and economic imperatives outweighed any exploratory wanderlust, or militaristic motivations.

Synergy Goes to War

A word is in order here about resurrecting perhaps the oldest of the prime mover theories about human evolution (as noted in Chapter 7). Warfare is, of course, a brutal and destructive business; the days are long gone when war was glorified as a noble pursuit. Nevertheless, war-making behaviors utilize many different kinds of functional synergy – synergies of scale, a division/combination of labor, functional complementarities, facilitation effects, human-tool symbioses, risk and cost-sharing, and more. It's not surprising that collective violence has been an important instrument for accomplishing the collective purposes of human groups historically.[100]

The thesis here is that our propensity for using collective violence as a survival tool has ancient roots, going back even to the australopithecines. It did not arise only with the emergence of humankind or modern civilization, and it endures because it continues to be perceived as being cost-effective – or at least effective at whatever cost. Nor are we unique in this regard. As I document elsewhere, collective violence is commonplace in the natural world, and rudimentary analogues of raiding and "warfare" have been observed alike in bacteria, social insects, and in many other animal species.[101]

Therefore, it seems likely that coercion played a significant part in the spread of modern humankind out of Africa and around the world, although it would be wrong to treat it as a prime-mover. This also conforms to the principle of parsimony. Collective aggression in the Upper Paleolithic represented an incremental adaptive change, the utilization of a synergistic behavioral pattern that had been a major part of the hominin behavioral repertoire for at least three million years.

Collective violence is, after all, a "tool" for attaining various ends; it is not an end in itself. The *casus belli* between evolving hominin populations were very likely shaped by many factors, including (perversely) our very success in reproducing and expanding our numbers.

Violent conflicts are more likely to occur in constricted environments, or when resource limits are reached. They are more likely when two populations are in direct competition for vital resources and there is little basis for cooperation – a zero sum game. Even then, bloodshed is not inevitable. The odds of violence are almost always influenced by a more or less explicit calculus of costs and benefits – and risks. A shorthand slogan for this calculus is, again, the "balance of power" (or, more to the point, an "imbalance of power"). However, this venerable concept implies a many-faceted analytical process, not a narrow statistical exercise, and it is conditioned by an array of other influences as well.

The Synergistic Ape

To sum up, there have been two major themes in human evolution. The first theme is that synergies of various kinds played a key role in the process. The economic incentives for the changes that occurred over time included synergies of scale, functional complementarities, novel combinations of labor, cost- and risk-sharing, and much more. Furthermore, each of these synergies supported and strengthened one another, and this in turn begat more synergy. For instance, the controlled use of fire was a common good that provided many survival-related benefits, but it also required a combination of labor for gathering fuel, for fire tending, for transporting hot coals from one campsite to another and, eventually, for fire making. This also necessitated a greater degree of social interdependence and cooperation, and in time it facilitated the development of cooking – another synergistic technology.

In other words, there was no prime mover or megamutation that suddenly allowed slow-witted hominins to think, and talk, and invent. Rather, an accumulation of many small inventions, both cultural and biological, ultimately produced a synergistic new package. A major threshold was crossed, rather like the achievement of fully-competent flight skills in evolving birds.[102] This synergistic "package" included efficient bipedalism, highly manipulative hands, larger and more sophisticated brains, numerous tools and technologies, elaborate patterns of social cooperation, organized collective activities, and a unique capacity to accumulate, use, and communicate cultural information. How do we know it was a synergistic package? Just apply the synergy-minus-one test (see Chapter 4). Take away bipedalism, or our dexterous forelimbs, or language, or even such important cultural attainments as fire. There is not a single one of these distinctive traits that we could do without.[103]

To put this evolutionary transformation into perspective, consider the fact that chimpanzees can do many of the things we are able to do, in a rudimentary way – walk bipedally, make tools, hunt in groups, communicate orally, make "war," even invent new techniques for food-getting and other needs.[104] Captive chimpanzees can also be trained to use symbols to converse and can even produce "art." The difference is that we can do all these things far better, and we do them cooperatively. The result is a difference in kind. Our synergistic genetic/cultural package is the cumulative result of an incremental, interactive and, to a large extent, purposeful evolutionary process – an entrepreneurial process.

Synergistic Selection

Accordingly, the second common thread in human evolution has been a dynamic of Synergistic Selection at the behavioral and cultural level. We progressively invented our species – although there was obviously no premeditated plan. The striking biological differences between humans and chimpanzees, or even humans and the Miocene midgets of 5–6 million years ago, are the result of a multi-million-year process in which new forms of cooperative behavior, new technologies, and new synergies, were the pacemakers. And each major new invention redefined the context of our evolution – both the selective biases and the ecological opportunities and threats. To repeat a punch line from Chapter 1, social cooperation has been the key to our evolution, and synergy is the reason why we cooperate. We have, in effect, invented ourselves, and synergy was the driver.

This cooperative, self-making hypothesis is not as radical or as new as it may sound. In fact, it highlights what is understated (or implicit) in many other theories. In *The Descent of Man*, Darwin himself suggested that learned behaviors – including new food procurement strategies, the invention of tools, and new forms of social cooperation – played a significant part in human evolution. (Recall that Darwin also embraced Lamarckian influences.) Likewise, in the Man the Hunter scenario of the 1960s, the transformative role of a behavioral innovation was stressed (see Chapter 7). The food-sharing and nuclear family scenarios of the 1970s also assumed that behavioral changes were key factors. Similar proposals about behavioral inventions and cooperation in human evolution – from allomothering to cooking – have been made by a number of other theorists more recently.[105]

Superorganisms

Finally, it should be emphasized that, from the outset, our remote hominin ancestors evolved in interdependent, cooperating groups – superorganisms. Competition via cooperation played a decisive role. For much of this multi-million-year journey, moreover, our ancestors followed an egalitarian pathway as defined by David Queller; they lived in functionally integrated social and economic units shaped by Synergistic Selection, even though many of the participants were unrelated and reproductively independent.[106] These ancestral hominin societies were, in effect, coalitions of families that were highly cooperative and benefited collectively from a skill-based, consensual leadership.[107] Mutualistic collaboration was the key to their success. Tomasello refers to it as the "interdependence hypothesis."[108]

And this "whole" – this collective survival enterprise – was held together by a combination of public goods (in important respects they were all in the same rowboat) and by corporate goods that were divided and shared among the members (again, recall the model in Chapter 3). The major transition to *Homo sapiens* was also distinctive in having an evolutionary unit that is functionally (culturally) defined, rather than genetically based, and it was, and is, sustained by an information system that allows for data to be distributed *ad libitum*, as well as being readily modified by cultural means and (more recently) even stored externally for use as necessary. In all of these ways, our ancestors invented a new survival strategy, which resulted in a major transition in evolution. We are at once the Synergistic Ape and the Self-Made Man.

---

[1] Indeed, as more fossil evidence has accumulated, some hotly contested distinctions have begun to blur. (For more on this point, see the outtake at my website.)

[2] http://www.nytimes.com/2015/09/11/science/south-africa-fossils-new-species-human-ancestor-homo-naledi.html?smprod=nytcore-ipad&smid=nytcore-ipad-share
New analyses indicate that the small-brained *H.* naledi may have co-existed with early humans 250,000 years ago.

[3] Asfaw *et al.* 1999. See also the report on meat-eating in De Heinzelin *et al.* 1999.

[4] Harmand *et al.* 2015; see also Alemseged *et al.* 2006; McPherron *et al.* 2010. (For more on this matter, see the outtake at my website.)

[5] See especially B. Campbell 1985, pp. 396-398.

[6] Dunbar 1996; 1998; 2001, 2009; Ambrose 2001; Sewall 2015.

[7] Foley 1995; Lewin 1993; Rodman and McHenry 1980; McHenry 1992; Steudel 1994, 1996; Klein 1999; Wolpoff 1999a; Pontzer *et al.* 2010.

[8] See Washburn and Lancaster 1968; Blumenschine 1987; Shipman and Walker 1989; Wrangham and Peterson 1996; Stanford 1999; Wolpoff 1999a,b; Wood and Collard 1999; Klein 1999. Wilkins *et al.* (2012) report evidence for hunting with hafted stone spears at least 500,000 years ago. See also Sahle *et al.* 2013.

[9] See especially deMenocal 2004, 2011.

[10] By contrast, the more "robust" hominins of that era, with a distinctively different anatomy and dentition, may have deployed a feeding strategy that relied on a broad range of lower-quality plant foods.

[11] Well-crafted, light throwing spears have been dated back at least 400,000 years. See Thieme 1997; Ambrose 2001; Stringer 2012. Jabbing spears that could seriously wound an animal were no doubt used much earlier.

[12] See Pickering 2013.

[13] Leonard and Robertson 1994, 1997; Leonard *et al.* 2007; also R.D.Martin 1981; Martin and MacLarnon 1985. The so-called expensive tissue hypothesis is based on the fact that the human brain represents about 2-3% of our total body mass yet consumes some 20% of our energy budget and even more in a developing infant. (For more on this, see the outtake at my website.)

[14] Aiello and Key 2002; see also Aiello and Wheeler 1995.

[15] Bramble and Lieberman 2004. (For more on this,see the outtake at my website.)

[16] The work of Peter Ungar and his colleagues suggests that there was also omnivory and a high degree of local opportunism. See Ungar and Sponheimer 2011; also, Ungar *et al.*

2006. There is even possible evidence of respiratory tract changes to accommodate a more strenuous life-style. See Trinkaus 1987.

[17] Newman 1970; also, Wheeler 1985, 1991. Discussions of sweating and hair loss as major adaptive changes in evolving humans can also be found in Foley (1995) and Wolpoff (1999a), among others.

[18] See Hrdy 2000, 2009; also, Burkart *et al.* 2009; 2014; also, the analysis of Kramer and Otárola-Castillo 2015. Also, see the "grandmother hypothesis" advanced by Kristin Hawkes (2003) and the three-generation model of Kaplan *et al.* 2009. In an important comparison across 78 primate species and 65 carnivore species, Caroline Schuppli *et al.* (2016) found that access to a more complex and stable foraging niche was highly correlated with a long childhood and parental food provisioning.

[19] Meet the Alloparents, *Natural History Magazine,* April 2009 http://www.naturalhistorymag.com/htmlsite/0409/0409_feature.pdf

[20] Darwin 1874/1871. See also the discussion in Wrangham and Carmody 2010.

[21] An older analysis, but still useful, is Clark and Harris 1985. A more recent analysis, based on archeological evidence for an upsurge in natural fires, can be found in Parker *et al.* 2016.

[22] Bellomo 1994; B. Campbell 1985; Kingdon 1993. Kingdon also points out that fire was too commonplace and too powerful in its effects to have been overlooked as a potential tool until only a few hundred thousand years ago.

[23] Wrangham *et al.* 1999; Wrangham 2009. However, Gowlett and Wrangham (2013) use a more conservative figure of 1.5 million years, while Stahlschmidt, *et al.* (2015) question the methodologies used in many of the findings and call for the use of new techniques for "microanalyses." But see also Parker *at al.* 2016.

[24] Wrangham et al. 1999; Wrangham 2009; Wrangham and Carmody 2010; Gowlett and Wrangham 2013; Smith et al. 2015. (For an elaboration, see the outtake at my website.)

[25] Wrangham 2009; see also Fonseca-Azevedo and Herculano-Houzel 2012; also, Herculano-Houzel 2012. (For more details, see the outtake at my website.)

[26] Wrangham believes that cooking may have played a major role in changing the mating system in hominins toward stable pair-bonds (the nuclear family), leading in turn to such anatomical changes as the concealment of ovulation and the permanent receptivity of the females. Wrangham argues that this social change was not primarily related to male provisioning, as Lovejoy supposed, but to the defense by males of their hard-won food supplies. Wrangham calls it the "theft hypothesis."

[27] A lengthy letter to the editor of the journal *Science* by anthropologist Ralph Rowlett (1999, p. 741) reinforced this argument. He concluded: "[Hominins] clearly had the pyrotechnical ability to cook tubers at least as far back as 1.6 million years ago, even if further research must determine exactly what was cooking." Further unambiguous evidence tracing back at least one million years was reported by Berna *et al.* 2012. See also Growlett and Wrangham 2013.

[28] Klein 1999; Wolpoff 1999a.

[29] Pinker 2010; also, Henrich 2016, p. 65ff. The idea of collective intelligence has gained support recently across several disciplines. See especially Couzin 2007; Woolley *et al.* 2010; van Schaik and Burkart 2011; Whiten and Erdal 2012.

[30] See especially Boehm 1993, 1997, 1999; also, Richerson and Boyd 1992; Gintis *et al.* 2015. For a recent critique and debate about what has also been called "strong reciprocity theory," see Guala 2009 and numerous commentaries. See also Boyd and Richerson 2005; Richerson and Boyd 2005; also, the similar argument in Wilson and Wilson 2008.

[31] Indeed, we now know that social learning is widespread throughout the animal world, including even in cetaceans, fishes, songbirds and, of course, the primates. See the special issue of the *Philosophical Transactions of the Royal Society B* in 2011, especially Whiten *et al.* 2011.

[32] Gavrilets and Richerson (2017), in an important new paper, model the evolution of the capacity to internalize and follow costly norms. They view this as a crucial step in human evolution.

[33] Piaget 1932; Kohlberg 1981; Kohlberg *et al.* 1983; Tomasello 2001, 2005, 2009, 2014, 2016.

[34] Tomasello develops his "shared intentionality" hypothesis at length in a 2014 book. (For details, see the outtake at my website.)

[35] Henrich 2016 pp. 56ff.; p. 212, p. 250. He argues compellingly that "culture makes us smart." We rely on "a large body of locally adaptive, culturally transmitted information that no single individual, or even group, is smart enough to figure out in a lifetime" (p. 12).

[36] Boyd *et al.* (2013), in a major review, argue that technological innovation is typically an iterative, cooperative process. (For more, see the outtake at my website.)

[37] Boesch and Tomasello 1998; see also Tomasello 2001, 2009. (For more on this, see the outtake at my website.)

[38] Henrich 2016, p. 57. He also envisions a "Rubicon" where a cultural evolution became the "primary driver" of our evolution. However, one must be careful not to reify culture as being some sort of independent agency or "driving force" (p. 317). The cultural innovations adopted by our ancestors were, for the most part, driven by their economic/functional benefits in terms of earning a living. The process became cumulative only insofar as it proved effective as a survival strategy. See also Thompson *et al.* 2016.

[39] Chomsky 2004.

[40] Pinker 1994; also, Pinker and Bloom 1990. See also the detailed review in https://en.wikipedia.org/wiki/Origin_of_language (last modified 22 June 2015).

[41] This is discussed at length in Corning 2005; also, Corning 2007b.

[42] This aspect of language is discussed in some depth in Richerson and Boyd 2010; Richerson and Christiansen, eds. 2013; see also Tomasello 2005, 2008; Henrich 2016.

[43] Four insightful discussions of this approach are Jablonka 2002, Tomasello 2008, Sewall 2015, and Henrich 2016. There is also much of value in the recent book by Bickerton (2009), although I disagree that recruitment for confrontational scavenging was the driver.

[44] Sherman *et al.* 1991, 1992.

[45] Suzuki *et al.* 2016.

[46] The case for language as a functional imperative in evolving hominins is argued in depth by Dunbar 1996, 1998, 2001, 2009; also, Henrich 2016. See also Deacon 1997.

[47] Mithen 2006. Also important, most likely, were body language and facial expressions, as Darwin himself (1965/1873) documented in his book *The Expression of the Emotions in Man and Animals.*

[48] Snowdon 2001. Michael Corbalis (2003), Michael Tomasello (2008), and others have theorized that intentional gestures (body language) likely preceded the emergence of verbal language skills among our ancestors. (For more, see the outtake at my website.)

[49] The idea of self-domestication has been applied especially to our propensity to be obedient rule-followers. As Henrich (2016) puts it: "Natural selection shaped our psychology to make us docile, ashamed at norm violations, and adept at acquiring and internalizing social norms. This is the process of self-domestication" (p. 319). See also Leach 2003; Russell 2012.

[50] See especially Tomasello 2005, 2008; also, Mazda Y Farias-Virgens and Yevgeniya Sosnovskaya, on Self-Domestication and the Evolution of Human Language: http://berkeleysciencereview.com/self-domestication-and-the-evolution-of-human-language/ (accessed 12 January 2016).

[51] See especially Levinson and Dediu 2013, Dediu *et al.* 2013; also, Szathmáry 2002, 2015; Deacon 1997; Szathmáry and Számadó 2008; Bickerton and Szathmáry 2011; Tomasello 2008.

[52] See the full discussion of this issue, and the evidence, in Deacon 1997.

[53] See Holloway 1983b. Deacon (1997) also points us to the conclusions of endocast researchers Philip Tobias and Dean Falk that language adaptations can be discerned even in *H. habilis.*

[54] See especially Levinson and Dediu 2013, Dediu *et al.* 2013; also, MacLarnon and Hewitt 1999.

[55] See especially Goren-Inbar 2011; also, Henrich 2016.

[56] See especially Levinson and Dediu 2013, Dediu *et al.* 2013; also, Dor and Jablonka 2000. A similar argument is developed in Richerson and Boyd (2010). They point out that the very diversity and local specificity of human languages acts to confine communications to insiders and exclude those who are not members of a group, reinforcing ethnocentrism.

[57] Aristotle 1946/ca. 350 B.C.

[58] Corning 2017.

[59] de Waal 1982; Strum 1987.

[60] Particularly noteworthy is the work of Mark van Vugt and his colleagues who have reformulated the field of leadership within an evolutionary/functional and psychological perspective. See especially Van Vugt 2006; Van Vugt and Ahuja 2011; also, the review by White 2011. See also the major cross-species study by J. Smith *et al.* 2016; also, Kummer 1968; de Waal 1997; Conradt and Roper 2003. Distributed control provides another, still poorly understood model. See Gordon 2007.

[61] Boehm 1993, 1997, 1999; see also Brown 1991.

[62] Gintis *et al.* 2015; see also Van Vugt 2006; Van Vugt and Ahula 2011.

[63] See Corning 2017. The concept of "prestige" as a basis for leadership is a well-established theme in anthropology. As Henrich (2016) observes: "Across human societies, prestige is consistently associated with great skill, knowledge and success in activities or tasks people care about. This prestige status readily forms a foundation for leadership in egalitarian societies" (p. 118). Consensual leadership derives ultimately from the convergent self-interests of the leaders and followers. See also Cheng *et al.* 2013. For a critical analysis, see Chapais 2015.

[64] Gintis *et al.* 2015. See also Boehm 1996, 1997; Puurtinen and Mappes 2009; Richerson *et al.* 2016.

[65] Anthropologist Richard G. Klein (2000) is the strongest proponent of a "punctuational" and biological cause. (For more on Klein, see the outtake at my website.)

[66] Wolpoff (1999a) and Klein (1999) interpret the available fossils and artifacts in different ways. Wolpoff sees a multi-regional pattern of progressive changes. Klein sees a more unitary process with periods of relative stasis interrupted by punctuational changes. Both acknowledge that progressive changes were occurring; see also Ambrose 2001; Wilkins *et al.* 2012.

[67] On this point, see especially Mcbrearty and Brooks 2000.

[68] Wolpoff 1999a, p. 554. Wolpoff's views were supported by subsequent discoveries of more sophisticated artifacts with much earlier dates in diverse regions. See Brooks *et al.* 1995; Yellen *et al.* 1995; McBrearty and Brooks 2000; Kuhn *et al.* 2001; d'Errico and Stringer 2011. It is also probably safe to assume that many more artifacts have decayed or were buried under modern settlements.

[69] Klein 2000, p. 26; also, Klein 1999.

[70] Stringer 2012, p. 240.

[71] Wolpoff *et al.* 1984; Wolpoff 1999b. Others have been associated with Clark Howell and Christy Turner.

[72] Wolpoff 1999b; Wolpoff *et al.* 2000; Jurmain *et al.* 2008; Stringer 2012.

[73] See the online recording of the 2016 CARTA symposium on Ancient DNA and Human Evolution, https://carta.anthropogeny.org/node/156412 (accessed 29 May 2016).

[74] See especially Klein 1999, 2000; Ehrlich 2000; Semino *et al.* 2000; Bar-Yosef 2002; Stringer 2003, 2012; Mellars 2006; Liu *et al.* 2006; Roberts 2009. There is much variation in the estimates for inclusive dates. Between 150-350,000 years ago, is also mentioned.

[75] It seems possible that some initial migrations occurred earlier. This would account for the human remains found at Skhul, Israel, that date to about 100,000 years and the recent finding by Alan Thorne and Darren Curnoe (2006) that an Australian fossil human, known as Mungo 3, now appears to be about 60,000 years old. See also Tucci and Akey 2016.

[76] Stringer 2012, p. 253. In a recent (2015) talk, Stringer described the process as a "coalescence" – a gradual accretion that was "assembled piecemeal." See https://royalsociety.org/events/2015/10/major-transitions/

[77] Hublin *et al.* 2017.

[78] Lieberman 1998, pp. 85-97; see also Tattersall 1998; compare with Deacon 1997. (For more on the FOXP2 gene, see the outtake at my website.)

[79] Diamond 1995.

[80] Tattersall 2000, pp. 56-62.

[81] Klein 1999; see also the discussion in Wade 2006.

[82] Richerson *et al.* 2005; Mellars 2006; Potts 2012; Marean 2015.

[83] Boyd and Richerson 2009; Richerson *et al.* 2009; Boyd *et al.* 2011; Richerson *et al.* 2016; also, Henrich 2016.

[84] A number of other theorists hold that the emergence of a more advanced technological package was an important factor in the modern human diaspora – Richard Klein 1999, 2009, Luca Cavalli-Sforza *et al.* 1994, Jared Diamond 1997, Christopher Wills 1993, 1998, Paul Ehrlich 2000, Ian Tattersall 2000, Paul Mellars 2006, John Shea and Matthew Sisk 2010, Daniel Lieberman 2013; Joseph Henrich 2016, and others.

[85] Richerson *et al.* 2009; Boyd *et al.* 2011; Collard *et al.* 2013; Derex *et al.* 2013. Marean stresses the evidence for extensive exploitation of marine resources. See https://royalsociety.org/events/2015/10/major-transitions/

[86] Richerson 2013; Henrich 2016; Powell *et al.* 2009. For a case against a population growth explanation, see Collard *et al.* 2016. (For more on this, see the outtake at my website.)

[87] Salisbury 1962, 1973.

[88] On hafted hand axes, see especially Clark 1992; Wadley *et al.* 2009. On the recent discoveries suggesting that bows and arrows were used as early as 64,000 years ago, see Lombard and Phillipson 2010; also, Mellars 2006. Tattersall (2000) notes that the more advanced, hafted axes had about 10 times as much cutting edge per pound of material as the more primitive hand axes. (For more on this matter, see the outtake at my website.)

[89] The history of this theory, from Darwin to the present day, is reviewed in van der Dennen 1995, 1999. (Some of the many other works related to this subject are cited in an outtake at my website.)

[90] A starting point for the extensive literature on ethnocentrism and xenophobia is Reynolds *et al.* 1987. (For more on this, see the outtake at my website.)
[91] See Corning 2007a and citations therein. (For more on this controversy, see the outtake at my website.)
[92] See Lahr *et al.* 2016. The authors found that at least 10 of the 12 skeletons found at a site near Lake Turkana appear to have died from violence. Also, see the extensive evidence in pre-historic California reported in M. Allen *et al.* 2016.
[93] Simonti *et al.* 2016.
[94] See Corning 2007a.
[95] Diamond 1997, pp. 53-57.
[96] See especially Clark 1992; Farmer 1994; Ambrose 2001; Mellars 2006; Wadley *et al.* 2009; Lombard and Phillipson 2010; Shea and Sisk 2010; d'Errico and Stringer 2011; Lieberman 2013; Marean 2015.
[97] See Mellars and French 2011.
[98] Marean 2015.
[99] In a recent (2015) talk, Marean emphasized the role of dense and predictable food resources in stimulating territoriality and lethal intergroup competition. See https://royalsociety.org/events/2015/10/major-transitions/
[100] The many different forms of synergy that are associated with collective violence in the natural world are explored in my paper on "Synergy Goes to War" (Corning 2007a).
[101] (More citations can be found in an outtake at my website.)
[102] Robert Foley, in a 2015 symposium on the major transitions in human evolution, concluded that it is our massive capabilities and extraordinary impact that sets us apart. This transition was not a punctuational event, he said, but a combined effect of influences at many levels. See https://royalsociety.org/events/2015/10/major-transitions/
[103] The provocative theory of Paul Bingham (2000) should also be mentioned. (See the outtake on this at my website.)
[104] For an in-depth comparison, see Whiten 2011.
[105] These theorists include C. Loring Brace, Milford Wolpoff, Richard Klein, Ralph Holloway, Jonathan Kingdon, Robert Foley, Terrence Deacon, Michael Tomasello, Edward Wilson, Martin Nowak, Kevin Laland and John Odling Smee, Boyd and Richerson, Kim Sterelny, Curtis Marean, Herbert Gintis and his colleagues, Joe Henrich, and others as well. (For more on this, see the outtake at my website.)
[106] See especially Stoelhorst and Richerson 2013; Lieberman 2013; also, Bowles 2009; Bowles and Gintis 2011.
[107] Recent studies among contemporary hunter-gatherers reveal that these groups contain many unrelated individuals, and that they have extensive social networks – both internally and externally – with neighbors and trading partners. This suggests that human cooperation has deep roots and goes far beyond inclusive fitness (Chapter Two). See Hill *et al.* 2011; Apicella *et al.* 2012.
[108] Tomasello *et al.* 2012.

Chapter 9

The Rise of Complex Societies

The collective survival strategy that our remote ancestors evolved over the past five million years or so remains at the core of our modus operandi as a species down to the present day.

Most of us live in deeply interdependent "tribes" that are organized to pursue our basic survival and reproductive needs cooperatively. Whatever may be our perceptions, or our illusions, a complex modern society is, in essence, a collective survival enterprise. We depend for the satisfaction of our basic needs on an elaborate division (combination) of labor supported by an awesome and ever-growing repertoire of tools and technologies, some of which we owe to the inventiveness of long-ago ancestors. Although the course of human history has obviously been far from smooth, the dominant trend has been an expanding and thickening web of economic and cultural synergies – and Synergistic Selection. Let's take a brief look at this evolutionary dynamic.

The emergence of larger, more complex human societies during the Holocene Epoch, beginning around 12,000 years ago, was a multifaceted process involving a suite of major changes, the elements of which can perhaps be distilled into four categories – *Settlements, Surpluses, Specialization, and Size.* A synergistic combination of more permanent encampments, an abundance of agricultural (and other) resources, an elaboration of different tasks and roles, and rapid population growth combined to create the foundation for the vastly larger and more productive economic systems of today.[1]

However, these important cultural developments were also undergirded and supported by an array of important pre-adaptations.

Many years ago, the archeologist and science writer John Pfeiffer noted that the rise of complex societies seemed to be closely associated with what he referred to as evolutionary "hot spots."[2] These were locations that possessed a rich combination of needed resources – concentrations of large game animals (or other protein sources like fish or shellfish), an abundance of edible plant materials, ample supplies of fresh water, plenty of firewood, and – as the agricultural revolution gained momentum – such things as a favorable growing climate, fertile soils, irrigation water, well-developed trading patterns, defensible terrain and, of course, an array of technologies.

Jared Diamond, in his path breaking study *Guns, Germs and Steel*, highlighted many of the cultural elements that were also instrumental to this momentous shift, including genetically-manipulated cereal grains and pulses, domesticated sheep, goats, and draft animals, irrigation systems, an array of specialized tools for plowing, harvesting, threshing, and grinding grains, cooking implements, food storage containers, record-keeping, defensive walls, and more.[3]

A Synergistic Package

Taken together, these elements created a synergistic new economic package that allowed for a sharp break with our egalitarian hunter-gatherer heritage – although this transition also seems to have happened in fits and starts.[4] Some of the earliest agricultural experiments were very modest and were later abandoned, or were overrun by enemies. Others seem to have been more of a hybrid between agriculture and hunting and gathering. There is, in fact, no definitive explanation for why this major cultural change occurred, but it is likely to have involved some combination of population growth, constraints on traditional food resources and consequent food shortages, plus new opportunities for growing and harvesting grains and vegetables, and the domestication and herding of animals.[5]

In any case, this package of cultural improvements resulted over time in changing the basic structure of human societies, as well as the relationships among its members.[6] Ironically, the very factors that contributed to our economic progress as a species also created opportunities for economic exploitation, social inequality, and political conflict. In effect, the egalitarian social contract that had sustained our hominin ancestors for millions of years was undermined, and this resulted in a deep structural defect that has plagued modern human societies down to the present day.[7] (I will discuss this problem further below and will suggest some possible remedies in the final chapter.)

As various anthropologists have concluded, the shift toward hierarchical societies started even during the late Stone Age, when complex hunter-gatherer tribes such as the affluent Nootka and Tlingit of the Pacific Northwest and the Natufians of the Eastern Mediterranean, became sedentary and began to display more elaborate social divisions and disparities of wealth.[8] Permanent settlements had the advantage of eliminating the time and labor required for frequent migrations, but they also created opportunities for an accumulation of personal property. (Of course, they also created attractive targets for covetous neighbors.) Reliable surpluses and a sedentary life-style also provided the wherewithal for developing crafts specialists and, in time, political, military, and religious hierarchies.

Recent research and new analyses point to the conclusion that valuable material property – such as agricultural land, tools, jewelry, gold, and eventually money – coupled with the ability of a family to retain these valuables through inheritance practices, played a key role in generating extremes of wealth and poverty (and power) in emerging complex societies.[9]

The Chumash Example

One well-studied example of this transition is described in detail by the anthropologists Kent Flannery and Joyce Marcus in their in-depth study, *The Creation of Inequality.* For more than 5,000 years, the Chumash, a population of Native Americans inhabiting parts of the Southern California coast and the Channel Islands, had lived as nomadic foragers. They subsisted mainly on coastal marine resources, as well as acorn groves and piñon nuts. Their small bands were also politically egalitarian.[10]

However, a radical change occurred between 500 and 700 A.D., when the Chumash invented ocean-going plank canoes with a sophisticated caulking system that could hold up to 12 crewmen and a ton of cargo and travel as far out to sea as 65 miles. This important new technology now gave the Chumash access to an abundance of large, deep-water fish, like giant tuna and swordfish, and this enabled them to feed larger, settled communities of about 150–250 people.

This new abundance in turn supported the development of a crafts industry and an extensive pattern of regional trade in such things as seashells, finished beads, ornaments, and highly valued flint stones and animal hides. Over time, some of the Chumash canoe captains (who owned the vessels, or at least monopolized them) were able to amass significant personal wealth and prestige, as well as multiple wives, and were able to pass their wealth and status along to their sons, or sometimes to close relatives. (The captains also typically served as war leaders and ceremonial chiefs and extracted tribute from their followers.) As a result, the other Chumash tribesmen, including especially the craftsmen, became heavily dependent on the canoe captains and were socially subservient.

A Change in the "Social Logic"

Flannery and Marcus conclude that the structural/political shift to a hierarchical society among the Chumash was not dictated entirely by technological and economic developments. It was ultimately due to a change in what the authors refer to as the "social logic" of a society – the basic ideology that is used to justify various social practices, especially those relating to wealth, rank, and hereditary privilege. Stories are fabricated to rationalize a desired social pattern or outcome for the wealthy. (A modern-day equivalent is the obviously self-serving political mantra that billionaires are the prime "job creators" in our societies.) Flannery and Marcus point out that a hierarchical society can only arise after changes in the social logic undermine the traditional social structure (the reverse dominance hierarchy, in Boehm's term).[11] "Each escalation of inequality required some overcoming of resistance."[12] To underscore this point, the authors cite examples of evolving societies where ambition, personal achievement, and social prestige did not result in the emergence of extreme material inequality or a hereditary hierarchy.

Far more common, unfortunately, were economic transitions that allowed for various elites to become self-aggrandizing and self-perpetuating. This was evidently the case even in some of the earliest agricultural villages that began to appear roughly 10,000 years ago in Mesopotamia. These villages, numbering perhaps 200–300 people, typically comprised various extended families. Each family lived in a separate mud-brick residential compound with as many as 15–20 rooms and several individual family hearths. They also seem to have had an elaborate division (combination) of labor and engaged in extensive trade with other settlements, where they could exchange their abundant agricultural surpluses (some villages had two harvests a year) for a variety of items, notably including brides that helped to cement alliances between trading partners.[13]

Growing Economic Inequalities

In many of these villages, it seems there were marked economic inequalities between families. Some of the residential compounds were much more elaborate, with larger rooms, large interior courtyards, and more abundant private storage rooms, as well as an accumulation of what archeologists refer to as "sumptuary goods" – items that connote luxury rather than utility. These included such things as painted pottery, statuettes, jewelry, and the like. Whether or not these affluent families also exerted a leadership role in their villages is unknown. (These early settlements were typically defended by perimeter ditches, walls, and watchtowers, with hand held slings and rocks serving as defensive weapons.)

In any case, as larger agricultural societies with populations numbering in the thousands began to appear in Mesopotamia, the so-called Big Men and chiefs that emerged over time began to develop hierarchies of political control and tax systems (and systems of conscripted "corvée labor" for public works) backed by coercive force. The once tight-knit, closely cooperating small communities expanded into much larger, more impersonal population centers – perhaps encompassing several neighboring villages organized around markets and trade – with many different economic enterprises, many specialized roles, and many potential conflicts of interest.

When the farmers and craftsmen who produced the food and other valuable items obtained equivalent benefits in return, including protection from external enemies, internal law and order, and the wherewithal to acquire desirable products from other specialists, this new division (combination) of labor – and property – was likely to have been perceived as equitable. Many early civilizations, particularly when they were still young, seemed to enjoy a degree of shared affluence and internal peace.[14]

But all too often this changed over time as various leaders and economic elites became increasingly exploitative. As the anthropologist Bruce Trigger points out in his magisterial synthesis, *Understanding Early Civilizations*, "A defining feature of all early civilizations was the institutionalized appropriation by a small ruling group of most of the wealth produced by the lower classes."[15]

In other words, the reverse dominance hierarchies that had long ensured against great disparities of power (and wealth) in traditional hunter-gatherer societies had broken down.[16] Voluntary consent gave way to top-down coercive force, and the traditional pattern of informal social controls and conflict-resolution was replaced over time by formal law codes, religious edicts, aggressive policing, and harsh punishments. (Many of these states also began to build public temples, monuments, palaces, and, eventually, armories and treasuries – the trappings of hierarchical societies everywhere.)

The Uruk Example

This radically new social structure can be observed even in some of the first-generation states that appeared about 5,700 years ago in what is today Iraq and southern Iran. Uruk, for example, represented an amalgamation of numerous villages with an urban center that ultimately grew to cover some 1.5 square miles with perhaps 20,000 inhabitants. Uruk was sustained by its rich alluvial soils and a network of irrigation canals supported by a highly productive agricultural system. This led to an elaborate division/combination of labor, including various administrative specialists, many different kinds of craftsmen, personal servants, and even slaves.

But most significant, Uruk also became a hereditary kingdom, with four distinct levels of social ranking.[17] The traditional reverse dominance hierarchy had, in effect, reverted to something akin to the typical primate pattern, although it was, of course, much more elaborately structured. An institutionalized successor to the traditional, egalitarian hunter gatherer political model – modern democracy – would not (formally) arise until 3,000 years later in classical Athens.[18]

Theories of "Civilization" and the State

Theorizing about the rise of large scale civilization and the emergence of the political machinery of the "state" over the past 6,000 years can be traced back at least to Plato's great dialogue, the *Republic*, and, more recently (and in more detail) to Herbert Spencer in the nineteenth century. Spencer posited a dualistic process of technological innovations and the elaboration of a division of labor (internal cooperation) coupled with warfare between societies (external conflict), although he also viewed Malthusian population pressures as a "proximate cause" and increased energy production as a key enabler.[19] "As societies progress in size and structure, they work on one another, now by their war-struggles and now by their industrial intercourse, profound metamorphoses."[20]

Unfortunately, Spencer's dualistic theory suffered a fate that was similar to the rejection of Lamarckism in biology. The rise of the social sciences as formal academic disciplines in the early years of the twentieth century was accompanied by a pronounced ideological tilt. Socialism was very much in the air in those days, and Spencer's early flirtation with what later came to be known as social Darwinism and his emphasis on the role of warfare in societal evolution became repugnant to many social scientists of that era. Spencer's encyclopedic, multi-volume theoretical edifice eventually came to be viewed as politically toxic and was banished from academic curriculums. "Who now reads Spencer?" intoned the prominent sociologist of that period, Talcott Parsons. "Spencer is dead."[21]

What ultimately replaced Spencer's complex dualism was a variety of prime mover theories. Perhaps the most influential of these theories was Karl Marx's deterministic vision of an "iron law" of history – a dialectical interaction between economic development and class conflict, with capitalism and private property as the ultimate villains and the state as a "handmaiden" of the capitalist ruling class.

As the twentieth century progressed, a number of other prime mover theories were also proposed. One of the earliest and best known of these was anthropologist V. Gordon Childe's thesis that the emergence of complex, large-scale societies could be attributed, at bottom, to agricultural surpluses.[22] His colleague Karl Wittfogel advanced a variation on this theme that was known as the "hydraulic hypothesis." Wittfogel postulated that the development of large-scale irrigation systems in Mesopotamia and China gave rise to bureaucratic controls, political stratification, and, in time, to what he termed "oriental despotism."[23]

Population Pressures

Another supposed prime mover – a resuscitated Spencerian theme without attribution – was population pressures. This theory was promoted by several theorists, but its most visible champion was the anthropologist Mark Nathan Cohen in a closely-reasoned, book-length argument. Characterizing population growth as the "cause of human progress," Cohen asserted that population pressure has been an "inherent" and "continuous" causal agency in cultural evolution over time.[24]

Among the various objections that have been raised to this theory is the fact that there are cases in which population pressures were mitigated by increased trade or an intensification of subsistence technologies, or even population control measures. The prior question is why populations can grow in some circumstances and not in others. It is not a given. In fact, there are numerous cases in which chiefdoms and states failed to emerge from a circumscribed context – when the environmental vice was in fact too tight for further expansion. Conversely, there is a very long list of unstable states that ultimately collapsed.[25]

Another resurrected Spencerian theme – improvements in energy production – was elevated to a prime mover theory by the anthropologist Leslie White in the 1950s. His "basic law of [cultural] evolution," in his words, was that "culture advances as the amount of energy harnessed per capita per year increases, or as the efficiency or economy of the means of controlling energy is increased, or both."[26] We cannot control the course of cultural change, White concluded, "but we can learn to predict it."[27]

Warfare as the Prime Mover

Perhaps the most strident of the modern-day prime mover theories about the rise of civilization is our old friend the warfare hypothesis. The evolutionary biologist Richard Alexander, for instance, developed a Darwinian, inclusive fitness explanation. He characterized war as a form of reproductive competition by other means and invoked the idea of an autocatalytic arms race. "At some point in our history the actual function of human groups – their significance for their individual members – was protection from the predatory effects of other human groups.... I am suggesting that all other adaptations associated with group living, such as cooperation in agriculture, fishing or industry, are secondary..."[28]

Another proponent of the warfare school of cultural evolution was the anthropologist Robert Carneiro. His theory was a bit more subtle (it relied on a functional argument rather than a presumed instinctual urge), but it too was monolithic. "Force, and not enlightened self-interest, is the mechanism by which political evolution has led, step by step, from autonomous villages to states." Although state-level political systems were invented independently several times, warfare was in every case the prime mover, Carneiro claimed. To support his thesis, he examined 21 instances of state development, ranging in time from 3000 B.C., to the nineteenth century A.D., and found that coercive force was a factor in every case and that outright conquest was involved in about half of them.[29] He blamed this dynamic ultimately on Spencerian population pressures – what he called "environmental circumscription."[30]

"The Fires of War"

An updated, more elaborate version of this theory has recently been advanced by ecologist Peter Turchin and his colleagues.[31] Turchin's thesis is that complex human societies were "forged in the fires of war." He systematically examined the historical records related to some 60 empires that have evolved over the past 5,000 years and found that 90 percent of them were located within or adjacent to arid steppe areas, or borderland steppe frontiers. Warfare in these areas was endemic, and the rise of complex states and early empires was closely associated with intense lethal competition between nomadic pastoralists and settled farmers, groups that were economically and culturally very different from one another.

Turchin stresses that larger numbers, coupled with technological advances like horse-drawn chariots and mounted archers with composite bows, played an important part in this competitive dynamic, and he sees this as being consistent with multi-level (group) selection theory in evolutionary biology. He notes that genocidal wars have been much more common between culturally distinct groups. Turchin also believes that hierarchical organization was necessary for the successful execution of large-scale warfare, and that this accelerated the rise of hierarchical political systems. Turchin argues that there is a consistent pattern in this historical trend, and that it provides compelling evidence for a war-related explanation of civilization and the state.

It is unquestionably true that organized warfare has been a major influence in the evolution of complex societies. Just as (I believe) collective violence played an important part in our emergence as a species (recall Chapters 7 and 8), it has also been deeply implicated in the evolution of larger civilizations. However, warfare is a complex phenomenon with many potential causes and many different consequences. Wars cannot simply be treated as the expression of an instinctual urge or uncontrollable population pressures, much less a mindless competitive dynamic. There are too many exceptions and too many problems with any monolithic theory.

For instance, why is it that some quite warlike societies – like the Yanomamö of Venezuela or the Dani of New Guinea – did not evolve into nation states? Why did some societies achieve statehood and then subsequently collapse or even disappear? It seems that, in some cases, climate change was the principal villain. And why did the first pristine states appear during a very small slice of time in the broader epic of evolution, within a few thousand years of one another at most? Nor is warfare always correlated with population pressures. Again, a prior question is why do some populations grow while others do not? There is even evidence of cases where population pressures and warfare were the result of economic and political integration rather than the reverse, with the proximate cause being ambitious (and/or paranoid) monarchs making war on one another, with popular support.[32] Because wars can be very risky, and costly, the central question is, what are the perceived (net) benefits? The potential payoffs must have been key drivers (as discussed in Chapter 8).

Economic Determinism Revisited

Indeed, Turchin's modernized warfare hypothesis has been challenged by a revitalized and more elaborate version of Childe's agricultural hypothesis. Economists John Gowdy and Lisi Krall assert that the underlying cause of our "ultrasociality" and the rise of complex modern states was the agricultural revolution and the economic benefits derived from it.[33] They stress the increasing economic returns (synergies) from large-scale agricultural production, which supported larger groups and a more complex division of labor. This in turn created political "management" challenges, and, ultimately, resulted in a radical change in the social structure of these expanding societies.

Agriculture certainly created new opportunities for resource exploitation and trade, but it also greatly intensified competition between groups. Population growth and an "expansionary dynamic" (as Gowdy and Krall call it) resulted from these economic developments, not the other way around.[34]

From Tin Pans to Water Pumps

There is one other frequently touted candidate for prime mover in cultural evolution that should also be mentioned – technology itself. Surely, nobody would dispute the fact that technology has played a major part in shaping modern societies over the past 10,000 years, with synergies that are readily quantifiable. The economic synergies that derive from new technologies can be seen in microcosm in the story of the great California Gold Rush. Over a five-year period, from 1848-1853, gold-mining techniques, in effect, recapitulated our entire technological phylogeny up to that time.[35]

Contrary to the mythology that has grown up around this renowned historical episode, most of the California gold mining activity was not done by individual prospectors – the legendary "sourdoughs" wading in mountain streams with tin pans. Within the first year, individual panning was largely supplanted by three-man teams using shovels and "rocker boxes," an innovation that increased the quantity of material that could be processed in a day from ten or fifteen buckets to more than 100 buckets, or at least twice as much per man. Shortly thereafter the wooden sluice made its appearance. Although it required six- to eight-man teams, a sluice could handle 400–500 buckets of material per day, or about twice again as much per man as the rocker box. Finally, when hydraulic mining was introduced in 1853, teams of 25 or more men were required to process the materials and manage the water pumps, hoses, wagons, etc., that were utilized to blast away the faces of entire hillsides. Meanwhile, the amount of material processed daily jumped to 100 tons or more. (This episode also illustrates the fact that almost every advance in technology creates a new imperative for social cooperation and organization.)

It's important to emphasize, however, that technology is not some exogenous, monolithic "force" or "mechanism." It's another example of an umbrella concept that embraces an immense array of synergistic phenomena. Some technologies involve simply deploying specialized knowledge and skills, like the use of dung as a fertilizer, or rotating crops. Others involve the manipulation of natural objects, like the selective breeding of plants and animals, or the diversion of water for the irrigation of crops. Physical structures like dams, walls, fences, and weirs are also important technologies and have played a significant role in human evolution. But most important, every new technology is embedded in a specific natural and cultural environment; it's catalyzed and thrives within a given economic, social and political context. The causes are often very complex – and synergistic. As Matt Ridley points out in his insightful book about how prosperity evolves, a major technological innovation is typically a "collective phenomenon."[36]

"Down with Prime Movers!"

Technological innovations have obviously been important drivers in our recent cultural evolution. But so were all the other factors that were singled out in the theories cited above – agricultural surpluses, population growth, warfare and conquests, and even fossil fuels. For example, the Industrial Revolution in England in the eighteenth and nineteenth centuries was literally powered by the discovery of abundant underground sources of coal coupled with the development of new mining and processing techniques that made it possible to extract coal much more cheaply and consume it more efficiently.[37] (More details about the Industrial Revolution can be found in an outtake at my website www.complexsystems.org.)

Perhaps Herbert Spencer's dualistic theory – which emphasized the interplay between internal economic development and external competition and conflict is closer to the right answer in this theoretical debate. As biologist David Sloan Wilson has expressed it, "Almost every school of thought has a baby and a bathwater." The question is: "What's worth keeping and what's worth throwing out."

In a major, book-length critique of cultural evolution theory back in the 1970s, the prominent anthropologist Elman Service came to this emphatic conclusion: "Down with prime movers!"[38] There is no "magic formula" that will explain human evolution, he wrote. However, there is one common theme – a "common denominator" in all the theories described above, namely, synergy.

Synergy Goes to War (Again)

Consider, for example, the warfare hypothesis. Earlier (in Chapter 8) I pointed out that collective violence is commonplace in the natural world and almost always has a specific purpose: predation, defense against predators, the acquisition of needed resources (food patches, nest-sites, water supplies, raw materials, territories, even mates), and the defense of these resources against other groups and species.[39]

However, the occurrence of collective violence – in nature and human societies alike – is facilitated by synergies of various kinds, which shape the calculus of bioeconomic benefits, costs and risks. Synergy is a necessary (though not sufficient) causal agency in collective violence. Among other things, there can be (1) synergies of scale, (2) cost and risk sharing, (3) a division (combination) of labor, (4) functional complementarities, (5) information sharing and collective intelligence, and, not least (6) tool and technology "symbioses." Although there are exceptions (and some significant qualifiers), collective violence is, by and large, an evolved, synergy-driven instrumentality, in nature and human societies alike.

Yet if warfare has had a major influence on our cultural evolution, it is clearly insufficient as an all-purpose explanation for the evolution of complex modern societies. We should move beyond our tendency to simplify the complexities of cultural evolution and resist the impulse to seek simplistic explanations – or scapegoats. Synergy is, above all, a concept that compels us to look for packages of interacting causal influences. (Indeed, the influence of political, legal, linguistic, even religious elements in shaping the process of state formation has been highlighted by various social scientists of late.)[40] The Synergism Hypothesis provides a theoretical bridge that can connect and integrate the various prime mover theories. In sum, the evolution of larger, more complex societies over time has been driven by a proximate dynamic of Synergistic Selection.[41]

An Autocatalytic Process?

From an evolutionary perspective, the convulsive growth and radical transformation of human societies over the past 250 years (or about twelve human generations) has been nothing short of phenomenal, and unprecedented. It seems as if the gradual process of cultural innovation and technological development that characterized human evolution for millions of years has now become autocatalytic; a vast, interactive ratchet effect is at work as new technologies feed on one another. The list of our major inventions to date is astonishing: steam power, railroads, steamships, electricity (and a plethora of electrical tools and appliances), telegraphs and telephones, automobiles, farm tractors (and sophisticated farm machinery), aircraft and helicopters, nuclear power, radio and television, container ships, space rockets, computers, the Internet, factory automation, robots, drones, and much more. What has facilitated this process is, of course, an economic (and political) system that offers large incentives for creativity and entrepreneurship, and mass markets that will reward these initiatives – in other words, a proximate dynamic of Synergistic Selection. This is the positive side of modern capitalism.

Among the many consequences of this seismic technological shift is the emergence of a vast global society that is a marvel to behold. Over the course of the past half century, we have become almost completely interconnected and increasingly interdependent in terms of food production, resource acquisition, manufacturing, transportation, communications, education, health care, and much more.

Consider, for instance, the transformative role of container ships. After shipping containers were introduced in the 1950s, the time required for overseas transport plunged by about 85 percent and the cost per ton declined by 35 percent. Sixty years later, our global container ship infrastructure is valued at $4 trillion. It includes about 450 ports and some 5,000 thousand huge container ships that (currently) move more than 1.6 billion metric tons of cargo every year, roughly five times as much as the 330 million tons in 1950.[42]

A Population Explosion

Perhaps the most significant outcome of our collective inventiveness and technological prowess is that it has subsidized a huge population explosion. At the end of the Great Famine and Black Death in 1350, the global human population was estimated to be about 370 million. Since then, the population has increased at an accelerating rate. It reached 1 billion in 1804, 2 billion in 1927, 4 billion in 1974, and 7 billion in 2011-2012. Although the rate of increase is now slowing, it is expected that the global population will grow by another 2–3 billion by 2050.[43] Humanity is reaching a climax phase, not unlike what has happened to other species. We are about to hit a ceiling at the very time when major climate changes are already underway. We are confronting the prospect of another major transition in our evolution.

What is most troubling – no, alarming – about our current predicament is that we do not have the cybernetic/political machinery to deal with this emerging crisis. As noted earlier, there is a deep and ominous structural flaw in modern societies that did not exist among our hunter-gatherer ancestors. The traditional reverse dominance hierarchy in humankind has devolved into various forms of exploitative hierarchical systems, for the most part. In the next chapter, I will focus on the existential challenges we face going forward and will suggest three urgently-needed societal changes.

---

[1] There is a debate among social scientists about how to define social complexity that is comparable to the terminological debate among evolutionary biologists. A synthetic approach recently suggested by a group of social scientists included such factors as population size, territory size, settlement density, economic specialization, trade

networks, management mechanisms, the number of levels of hierarchical control, and more. See Richerson and Christiansen, eds. 2013, pp. 87-116.
[2] Pfeiffer 1977.
[3] Diamond 1997. For a recent discussion of the role of technology in human evolution, see Boyd *et al.* 2013.
[4] See especially Bogucki 1999; Mithen 2003.
[5] Richerson *et al.* (2001) argue that agriculture became "compulsory" in the Holocene as a result of a "competitive ratchet" driven by competition between groups in the context of population pressures. See also Sterelny 2013; also, the analysis of Winterhalder and Kennett (2009) that utilizes the economic concepts of risk, discounting, economies of scale, and transaction costs. Social networks and more elaborate patterns of exchange may also have played a significant role. See Apicella *et al.* 2012.
[6] See especially the analysis in Gowdy and Krall 2014.
[7] An analysis of this transition can be found in Flannary and Marcus 2012. What seems to hold the social contract together in complex societies, despite the extreme differences in wealth and poverty, is some combination of economic interdependence, orderly exchange (markets), the need for a common defense, social and cultural/communal bonds, coercion (there is no available alternative) and policing.
[8] See especially Fagan 1998; also, Flannery and Marcus 2012.
[9] See the special issue of *Current Anthropology* in 2010 devoted to this subject. (For citations, see the outtake at my website.) See also the analysis of the "endowment effect" in facilitating the evolution of private property in various species by Gintis 2007.
[10] Flannery and Marcus 2012, pp. 67-71. Their example was derived from an in-depth study by archeologist Jeanne Arnold and her co-workers. Many details also come from eye witness accounts by early eighteenth century Spanish colonists.
[11] Boehm 1993, 1999. Recall that Boehm's term refers to coalitions that actively contain and suppress aggressive individuals. See also Brown 1991.
[12] Flannery and Marcus 2012, p. 563.
[13] *Ibid.* Recent DNA studies have revealed that the earliest farming communities arose independently. Only much later did they interact (and interbreed). See especially http://www.nytimes.com/2016/10/18/science/ancient-farmers-archaeology-dna.html?action=click&contentCollection=science®ion=rank&module =package&version=highlights&contentPlacement=1&pgtype=sectionfront_ (accessed 18 October 2016).
[14] A notable example is the Indus River civilization known as the Harappans, from about 4500 to 4000 years ago, which evidently never had a centralized ruling class or extremes of wealth and poverty. See Maisels 1999.
[15] Trigger 2003, p. 375.
[16] Boehm 1993, 1999. Sterelny (2013) argues that a shift from reliance on "social capital" (skills) to "material capital" (land) played a key role. But so did the

transformation of the economic system from one that was based on collective action to one that was based on exchange and reciprocity among different specialists.

[17] See Flannery and Marcus 2012.

[18] For a brief history, see https://en.wikipedia.org/wiki/History_of_democracy (last modified 30 October 2105).

[19] Spencer 1852, 1897.

[20] Spencer 1897, vol. 1, 1 pp. 14-15.

[21] Parsons 1949/1937, p. 3.

[22] Childe 1951/1936.

[23] Wittfogel 1957.

[24] Cohen 1977.

[25] See especially Jared Diamond's (2005) book length treatment. Also, see Corning 2005, Ch. 7.

[26] White 1959, p. 56.

[27] White 1949, p. 39.

[28] Alexander 1987.

[29] Carneiro 1981; also, Carneiro 1978.

[30] Carneiro 1970. An example of this dynamic can be found in the history of the Zulu nation. (For citations, see the outtake at my website.)

[31] Turchin 2009, 2011, 2016; Turchin and Gavrilets 2009.

[32] See Corning 2005, Ch. 7; also, Corning 2007a.

[33] Gowdy and Krall 2014, 2015.

[34] A variation on this theme is physical chemist Ugo Bardi's (2014) theory that the rise of complex states was driven by a quest for mineral resources – gold, iron, bronze, etc. Yes, but first you must feed the population.

[35] This material was obtained personally from displays at the Gold Mining Museum, Angels Camp, California, September 2004.

[36] Ridley 2010, p. 4.

[37] See http://en.wikipedia.org/wiki/Industrial_Revolution (last modified 10 November 2014).

[38] Service 1971, p. 25.

[39] See also Corning 2007a.

[40] Political scientist Roger Masters (2008), in a systematic analysis of state formation, identified a number of important contributing factors, from a favorable environment to exceptional leaders, ethnic and religious commonalities, a common language, writing and record keeping, monetary systems, mutually beneficial economic activities, and, of course, military conquest. See also Richerson and Christiansen, eds. 2013.

[41] To be clear, this theory does not make a claim to explain every aspect of culture, and cultural change. It applies to functional improvements over time in the "means of production" and reproduction – the "collective survival enterprise." (For additional

discussion of this issue, see the outtake at my website.). Also, see Corning 1983, 2005; Plotkin 2010; Mesoudi 2011; Richerson *et al.* 2016.

[42] See https://en.wikipedia.org/wiki/Container_ship (last modified 19 July 2015); also see http://www.worldshipping.org/about-the-industry/liner-ships/container-vessel-fleet (accessed 24 July 2015); https://www.statista.com/topics/1367/container-shipping/ (accessed 26 March 2017).

[43] https://en.wikipedia.org/wiki/World_population (last modified 23 July 2015).

Chapter 10

The Next Major Transition

The conventional wisdom among biologists (and many others as well) until quite recently was that human evolution is now somehow "complete." Typical of this view was the widely quoted remark by the paleontologist/popularizer Stephen Jay Gould some years ago: "There's been no biological change in humans in 40,000 or 50,000 years. Everything we call culture and civilization we've built with the same body and brain."[1]

We now know this conceit is wrong on two counts – no three. In reality, the evolution of our species is still a work in progress – an unfinished symphony. For one thing, it's clear that our cultural heritage long predates the emergence of humankind, and that it has shaped the trajectory of our biological evolution (see Chapters 7 and 8). Second, and more important, it's increasingly evident that our species is continuing to evolve, both culturally and biologically. The evolutionary process may even be accelerating.[2] And this says nothing about the potentially revolutionary implications of genetic engineering.

For starters, each of us carries with us about 50 new genetic mutations that our parents did not have. In all, it is estimated that there have been more than one million new genetic variations in humankind since the rise of agriculture. But far more significant is the evidence compiled in the International Haplotype Map showing that there are marked genetic differences between contemporary human populations. In fact, there has been recent selection (within the past 40,000 years) across about 20 percent of our genes.[3] These biological changes include such obvious things as regional differences in skin pigmentation, thicker subcutaneous fat layers among cold climate populations, respiratory adaptations in high altitude peoples, adult lactose tolerance in places where cattle and milk-drinking are common, as well as an evolving resistance to various diseases (tuberculosis, small pox, malaria, etc.).[4]

Geneticists also suspect that evolutionary influences are still at work affecting changes in our body stature, our teeth, and perhaps even personality traits like mood, tolerance for stress, and reactions to various social conditions. Biologist Kevin Laland and his colleagues have identified eight specific gene sets – ranging in number from 2 to 31 – that are linked to the expression of different cultural traits – from the consumption of cooked foods to local climate adaptations.[5] Many more genetic changes may be occurring these days as modern humans (and many other species as well) increasingly shift to living in vast urban conglomerates. In short, gene-culture co-evolution is still ongoing.

The Anthropocene Epoch

Finally, it's now apparent that the growing tide of humanity is rapidly reshaping the entire biosphere and influencing the fate of our own and many other species as well. It seems appropriate to call this the Anthropocene Epoch (the age of human dominance) rather than the Holocene, as some geologists have proposed to do.[6] Whether you choose to characterize this dynamic as vicious or virtuous, there is a circle of causation at work here that will profoundly affect our destiny as a species. For better or worse, we are on the cusp of another major transition in evolution.

It seems that the law of unintended consequences has a large subcategory that could be called the law of unintended synergies. There should be a rule of thumb which says that for every positive synergy that humankind produces there will be some offsetting negative synergy as well. Ecologist Garrett Hardin, legendary for "The Tragedy of the Commons" (see below), is also known for his First Law of Ecology: "You cannot do only one thing."[7]

Economists typically treat such negative synergies as a production cost in their cost-benefit analyses, or preferably an "externality" (a cost for somebody else) that can be excluded from their calculus altogether. Industrial pollution and climate warming, with all their adverse consequences, are probably the most flagrant examples. But equally alarming is the rapid destruction of the world's topsoil (the U.N. projects that, at the current rate of depletion, it could be gone in 60 years),[8] not to mention the relentless contamination and depletion of precious fresh water supplies, the precipitous decline of marine resources (a major share of our food supply), the tragic loss of biodiversity in our ecosystems, the growing threat to our boreal forests, and the clear-cutting of irreplaceable rain forests. Or consider the recent drought in India, which threatened some 330 million of its people.[9] There will be many more such life-threatening droughts in the future. Perhaps most menacing is the evidence that sea levels may be rising at an accelerating rate due to climate warming, with potentially catastrophic consequences over the longer term for the world's coastal cities and their hundreds of millions of inhabitants.[10]

A Cul de Sac

It's not exactly news that these and other negative synergies are now beginning to overwhelm us, just as Walter Lippmann predicted so many years ago (recall his words in Chapter 1) and as many others have repeatedly warned since then. The awe inspiring technological niche that we have constructed for ourselves has become a cul de sac, and we now find ourselves in mortal peril. But there is no going back. We've lost the cultural know-how required to survive as foraging bands or hunter gatherers, not to mention the wild resources needed to sustain them.

While it's true that every species is always theoretically at risk of extinction, life is all the more precarious for our own high-impact, high-maintenance species, with a global population that is still growing and a survival strategy that is unsustainable. Humankind has already exceeded its viable long-term limit, yet we continue to pursue economic growth as if it were the Holy Grail – the all-purpose solution to our numerous problems. This strategy will no longer work. It can only hasten our ecological judgment day. To paraphrase an old saying, if growth can't go on forever in a finite world, it won't.[11]

A recent article in the *Proceedings of the National Academy of Sciences,* co-authored by biologist Paul Ehrlich (a leading environmental advocate who is famous for his 1968 best seller *The Population Bomb*) together with the environmental scientist John Harte, concluded that "To feed the world in 2050 will require a global revolution.... Anything less is a recipe for disaster."[12] Their alarming prognosis was based on a new report from the United Nations' Food and Agriculture Organization. The report estimates that at least 2 billion people world-wide (roughly one out of four) are currently going hungry or are malnourished and that a 70 percent increase in global food production will be needed to adequately feed the projected global population in 2050.[13]

Green Shoots?

An important two-hour documentary film about our current predicament, "Humanity from Space," which aired in 2015 on the public television network, ended on a positive note. After acknowledging that we cannot for much longer continue our present course (for one thing, our fossil fuel reserves will only last for the remainder of this century, even if burning them all would cause no harm to the environment), the film stressed that our unique inventiveness as a species and our growing interconnectedness are advantages that will ultimately enable us to solve our formidable problems.

To illustrate this optimistic conclusion, the film featured a pioneering "food factory" in Chicago that gets multiple harvests each year from vertically stacked indoor growing beds with LED grow lights. The film also highlighted a solar farm in the California desert, where thousands of mirrors are arrayed around a central tower that generates enough steam to produce electric power for 140,000 homes.

These and other green shoots are hopeful signs, but they could all be too little, too late. More to the point, our primary challenge is *not* about finding technological solutions. It's about how we govern ourselves – or fail to do so, as Lippmann warned so long ago and as Plato before him pointed out. Former President Obama, in his historic speech at the Hiroshima Memorial in 2016, warned that "Technological progress without equivalent progress in human institutions can doom us."[14]

In an article in the journal *Science* recently, biologists Toby Kiers and Stuart West noted that all the previous major transitions in evolution have depended on (1) an alignment of interests, (2) mutual dependence, and (3) effective ways to curtail internal competition, cheating, and free riding – in other words, self-governance for the common good.[15] We are not even close to achieving this trifecta in humankind. Our supreme political challenge going forward will be to create what could truly be called the ultimate superorganism. It would be the next major transition in the evolutionary process.

An "Engineering" Defect

As discussed in Chapter 9, complex modern societies suffer from a serious structural defect, a political "engineering" problem. Every biological system requires an effective sub-system of cybernetic communications and control – or governance for the good of the whole (recall the discussion in Chapter 6). But modern humankind lacks the consistent ability to ensure that individual, corporate and tribal/national interests adhere to the common good and the needs of posterity, in contrast with, say, leaf cutter ants (see Chapter 4). Modern human societies are at best "crude superorganisms," as Richerson and Boyd put it, political workarounds that are often hamstrung – or corrupt – and unable to act in the public interest.[16]

Needless to say, our emerging global superorganism is even more underdeveloped and dysfunctional. We are collectively on a path that cannot be sustained for much longer. Our growing interdependence – both positive and negative (think terrorism, refugees, pandemics, and offshore tax havens) – requires a new level of cooperation (and governance) to match the scale and inter-connectedness of our global survival enterprise. Moreover, the need for governance will become even more urgent as climate change increasingly disrupts the global environment and threatens the lives and well-being of many millions of people. We must reconstitute and greatly expand the social contract that

that was a key to our evolutionary success as a species and that sustained our ancestors for millions of years (see Chapters 7 and 8). To achieve the next major transition in evolution, we will need to change (1) our basic social values, (2) the actions of our vested and powerful economic interests, and (3) the role of government. This is obviously a very tall order. Where do we start?

The Collective Survival Enterprise

The overall challenge we face can be framed in terms of the fundamental purpose of a human society. All of us are – to repeat – participants in a multi-million-year-old collective survival enterprise. Survival and reproduction remains the basic, continuing, inescapable problem for every living organism, including humans, and it is this biological imperative that defines the ultimate priorities for every society.

It may come as a surprise to learn that the collective survival enterprise in humankind entails no less than fourteen distinct categories of "basic needs" – absolute requisites for the survival and reproduction of each individual, and of society as a whole over time.[17] Furthermore, we spend most of our daily lives involved in activities that are either directly or indirectly related to satisfying these needs, including (not least) earning a living and contributing in various ways to help sustain the collective survival enterprise.

These fourteen basic needs domains include a number of obvious categories, like adequate nutrition, fresh water, physical safety, physical health, mental health, and waste elimination, as well as some items that we may take for granted, like thermoregulation (which encompasses many different technologies, from clothing to blankets, fire wood, heating oil, and air conditioning). Our basic needs even include adequate sleep (about one-third of our lives), mobility, and healthy respiration, which can't always be assured these days. Perhaps least obvious but most important are the requirements for reproducing and nurturing the next generation. In short, our basic biological needs cut a very broad swath through our economy and our society. (These fourteen needs categories are discussed in more detail in my 2011 book, *The Fair Society*.)[18]

To repeat, the basic challenge for every human society is to provide for the survival and reproductive needs of its members. This is our prime directive. However, it is obvious that we are currently falling far short, and the situation is likely to get much worse. The next major transition in evolution will require a refocusing of our social values and actions so that they are more fully aligned with the basic purpose of the collective survival enterprise and our basic needs. In short, we must develop a new, biologically-oriented approach to our social relationships and our mutual obligations. I refer to it as a "biosocial contract" because it's derived from the emerging science of human nature and our documented biological needs.

A Basic Needs Guarantee

A reformulated social contract must start with a universal "basic needs guarantee." The case for this foundational principle is grounded in four key propositions: (1) our basic needs are increasingly well-understood and documented; (2) although our individual needs vary somewhat, in general they are equally shared by all of us; (3) we are dependent upon many others, and our economy as a whole, for the satisfaction of these needs; and (4) more or less severe harm will result if any of these needs is not satisfied.

The idea of providing everyone with a basic needs guarantee may seem radically new – a utopian moral aspiration, or perhaps warmed-over Marxism.[19] However, it's important to stress that this would not entail an open-ended commitment. And it is emphatically not about an equal share of the wealth. It refers to the fourteen domains of basic biological needs that were cited above. Our basic needs are not a vague theoretical abstraction, nor a matter of personal preference. They constitute a concrete but ultimately limited agenda, with measurable indicators for evaluating outcomes.

A basic needs guarantee also has strong public support. For instance, a famous (and much replicated) series of social experiments first conducted by political scientists Norman Frohlich and Joe Oppenheimer in the 1990s found that some 78 percent of the participants overall favored ensuring a basic economic "floor" for everyone.[20] A more recent public survey by researchers at Harvard University found that 47 percent of young people in the U.S. between the ages of 18 and 29 agree with the proposition that our basic necessities should be treated as "a right that government should provide to those who are unable to afford them."[21] There is also growing interest these days in the convergent idea of providing everyone with a "universal basic income." It's an old idea that has enlisted many prominent advocates over the years.[22]

The Right to Life

The argument for a basic needs guarantee also accords with the "right to life" principle. The philosopher John Locke in his *Two Treatises of Government* (1690) was the first "modern" theorist to assert the idea of self-evident human rights, including "life, liberty and estate [i.e., property]."[23] However, Locke stressed that our rights are not absolute. They must not interfere with the rights of others. Furthermore, Locke insisted, governments exist to protect these rights.

In the same spirit, Adam Smith in *The Theory of Moral Sentiments* emphasized the importance of doing justice, which he defined as not causing injury to others. "There can be no proper motive for hurting our neighbor."[24] The utilitarian philosopher Jeremy Bentham also qualified his signature "pain-pleasure" ethical principle by conceding that our freedom must be constrained by the rule that it "affects the interests of no other persons" besides the actor.[25] Modern-day libertarians, likewise, generally acknowledge that the exercise of our rights must not cause "harm" to anyone else. (See, for example, philosopher Robert Nozick's often-cited 1974 book, *Anarchy, State and Utopia.*)

The first public (political) assertion of a right to life is, of course, enshrined in the American Declaration of Independence (1776), and it has been invoked in many other contexts since then, including the United Nations' Universal Declaration of Human Rights (1948), the European Convention on Human Rights (1950), the U.N.'s International Covenant on Civil and Political Rights (1966) and the Convention on the Rights of the Child (1989), as well as the Basic Law for the Federal Republic of Germany (1949), the Indian Constitution (1950), and the Catholic Church's Charter of the Rights of the Family (1983). The right to life is also frequently used in public debates over such issues as capital punishment, euthanasia, and (in the U.S.) anti-abortion advocacy.[26]

But if the right to life is widely-recognized as a self-evident moral principle (although it's often dishonored in practice), it certainly does not end at birth; it extends throughout our lives. Moreover, it is a prerequisite for any other rights, including liberty and "the pursuit of happiness" (or property rights, for that matter). The right to life necessarily also implies a right to the means for life – the wherewithal. Otherwise this right is meaningless. And because almost all of us are dependent upon the collective survival enterprise to obtain the "goods and services" required for satisfying our personal needs, the right to life imposes upon society and its members a life-long mutual obligation to provide for one another's basic needs. This includes reproduction and the nurturance of the next generation.[27]

The "Fair Society" Model

One obvious objection to creating a basic needs guarantee is that it amounts to a give-away; it would invite free-riding. Where's the fairness in that? As Kiers and West noted in the *Science* article cited above, this could pose a serious obstacle; there must be effective measures to prevent cheating and free-riding.

In *The Fair Society*, I argued that social justice has three distinct aspects. Our basic needs must take priority, but it is also important to recognize the many differences in *merit* among us and to reward (or punish) them as appropriate. It is well documented that the principle of "just deserts" also plays a fundamental role in our social relationships. In addition, there must be reciprocity – an unequivocal commitment on the part of all of us (with some obvious exceptions) to help support the collective survival enterprise. We must all contribute a fair share toward balancing the scale of benefits and costs, for no society can long exist on a diet of altruism. Altruism is a means to a limited end (helping those in genuine need), not an end in itself. We must reciprocate for the benefits that we receive from society through such things as our labor, the taxes we pay, and public service.

Accordingly, the Fair Society model includes three distinct normative (and policy) precepts that must be bundled together and balanced in order to achieve a stable and relatively harmonious social order. It could be likened to a three-legged-stool; all three legs are equally important. A shorthand version of these precepts is *equality, equity,* and *reciprocity.* (They are discussed at length in *The Fair Society.*) To be specific:

> **(1) Goods and services must be distributed to each according to his or her basic needs (in this regard, there must be *equality*);**
>
> **(2) Surpluses beyond the provisioning of our basic needs must be distributed according to "merit" (there must also be *equity*);**
>
> **(3) In return, each of us is obligated to contribute to the collective survival enterprise proportionately in accordance with his or her ability (there must be *reciprocity*).**

Plato, in his great dialogue, the *Republic*, defined social justice as "giving every man his due." (The little-used subtitle of the *Republic* is "*Concerning Justice.*")[28] Yes, but what is a person's "due"? In the Fair Society model, social justice has three substantive elements: *equality*, *equity*, and *reciprocity.* First among these is equality with respect to the right to life and a basic needs guarantee, but all three of these principles are essential for the next major transition in evolution.

Going forward, a basic needs guarantee must become the moral foundation for every human society. It provides specific content for the Golden Rule and a shopping list for the Good Samaritan. It speaks to the fundamental purpose of the collective survival enterprise, and it represents perhaps our greatest ethical and political challenge. Equally important, a basic needs guarantee is an absolute prerequisite for achieving the level of social trust, harmony, and legitimacy that will be required to heal our deep social and political divisions and respond effectively to global warming and our growing environmental crisis.

Capitalism is Part of the Problem

But this will hardly suffice. The second major change in our modus operandi as a species must involve the values (and the outcomes) in our economic systems. Our ecological crisis has many contributing causes, but the root of the matter is modern capitalism – at once an ideology, an economic system, a bundle of technologies, and an elaborate superstructure of supportive institutions, laws and practices that have evolved over hundreds of years. Capitalism has the cardinal virtue of rewarding innovation, initiative and personal achievement, but it is grounded in a flawed set of assumptions about the nature and purpose of human societies and our implicit social contract; its core values are skewed.[29]

In the idealized capitalist model, an organized society is essentially a marketplace where goods and services are exchanged in arms-length transactions among autonomous purveyors who are independently pursuing their own self-interests. This model is in turn supported by the assumption that our motivations can be reduced to the efficient pursuit of our personal "tastes and preferences." We are all rational "utility maximizers" – or *Homo economicus* in the time-honored term. This is all for the best, or so it is claimed, because it will, on balance, produce the "greatest good for the greatest number" (to use the mantra of utilitarianism). A corollary of this assumption is that there should be an unrestrained right to private property and the accumulation of wealth, because (in theory) this will generate the capital required for further economic growth. More growth, in turn, will lead to still more wealth.[30]

The foundational expression of this model, quoted in virtually every introductory Economics 101 textbook, is Adam Smith's invisible hand metaphor. As Smith wrote in *The Wealth of Nations*, "man is...led by an invisible hand to promote an end which was no part of his intention. Nor is it always the worse for the society that it was not part of it. By pursuing his own interest, he frequently promotes that of the society more effectually than when he really intends to promote it.... In spite of their natural selfishness and rapacity...[men] are led by an invisible hand to...advance the interest of the society..."[31]

"Utopian Capitalism"

The classical economists who followed in Smith's footsteps embellished his core vision in various ways. One of the most important of these early theorists, Léon Walras, claimed that the market forces of supply and demand, if left alone, would work to ensure the efficient use of resources, full employment, and a "general equilibrium." In other words, competitive free markets can be depended upon to be self-organizing and self-correcting, and the profits that flow to the property owners – the capitalists – will generate the wherewithal for further growth and, ultimately, the general welfare. The modern economist Robert Solow summed up what has been called (sometimes derisively) "utopian capitalism" as a compound of "equilibrium, greed and rationality."[32]

The well-known senior economist Samuel Bowles, in his recent book-length critique and re-visioning of economic theory with the unassuming title *Microeconomics*, points out that capitalist doctrine offers "an odd utopia."[33] Its strongest claims are generally false; it is unable to make reliable predictions; it removes from its models many of the factors that shape real-world economies; it ignores the pervasive and inescapable influence of wealth and power in shaping how real economies work; and, not least, it's profoundly unfair. It systematically favors capital over labor, with results that are evident in our skewed economic statistics and widespread poverty. Senior economist John Gowdy candidly acknowledges that "Economic theory not only describes how resources are allocated, it provides a justification for wealth, poverty, and exploitation."[34]

The U.S. Example

Take, for example, the United States in 2010. In that year, the top one percent of the population held 35.5% of the wealth (now it's closer to

40%) while the top ten percent held 77.1% of the wealth. Likewise, the top one percent received 21% of the total annual income and the top ten percent received 49.2%. The remaining ninety percent had to split the other half. During that year, some 15–25 percent of the American population lived in more or less severe poverty (depending on which method you use to compute it), while 48 million people had no health insurance and about 50 million (including many millions of children) experienced some "food deprivation."[35] In the 2015 edition of the new Social Progress Index, the U.S. was ranked 16th among 133 nations, despite being first in GDP.

In other words, the "corporate goods" (see Chapter 3) that our society produces have been very inequitably divided. And, alas, the U.S. is not atypical. Economist Thomas Picketty, in his acclaimed and heavily-documented 2013 study *Capitalism in the Twenty-First Century*, concludes that extremes of wealth and poverty have been the rule in the modern world rather than the exception.[36] At bottom, this is a consequence of the serious structural problem in human societies that arose during the agricultural revolution, when our traditionally egalitarian small "tribes" expanded in size and complexity and became hierarchical and deeply inequitable (see Chapter 9).

To be sure, modern capitalism comes in many different sizes and shapes, from the millions of small mom-and-pop businesses with only one or a few workers to huge international conglomerates with hundreds of thousands of employees world-wide. But for every Google that provides a cornucopia of perks for its employees there are many others that are single-mindedly devoted to an iron triangle of mutually reinforcing values: (1) maximizing growth, (2) maximizing efficiency, and (3) maximizing profitability for the owners/managers and the shareholders.

A Rigged Game

In fact, a modern capitalist economy can become a rigged game that is far removed from the idealized market model – individual actors with equal power and resources who rationally pursue their self-interests with mutually beneficial (win-win) exchanges that result in efficient markets and optimal outcomes for all concerned. Among other things, the vast differences in wealth, power, and information between the actors exert a highly coercive influence in the marketplace, and in our political system. (For a classic critique, see David Korten's *When Corporations Rule the World.*)[37] Many different adjectives have been used to describe such market distortions: crony capitalism, klepto-capitalism, mafia capitalism, ersatz capitalism, casino capitalism, permissive capitalism, subsidized capitalism, and more.

The Rules of the Game

These perversions flout the very principles of social justice that I outlined above. This is why the second of the three key changes that will be necessary to achieve the next major transition in evolution must be a fundamental shift in our economic/business values and practices. Bill Gates Jr. (lately the richest man in the world and nowadays a prominent philanthropist) summed up the problem succinctly in a TV interview a few years back: "Markets only work for people who have money."

In reality, it's not abstract markets but people and their social values (and the rules under which they play the game) that shape the actions of business firms and the workings of the economy.[38] Capitalism mainly produces corporate goods, and these have been inequitably shared for the most part, often without regard for life-and-death externalities, not to mention damage to the environment.

Pope Francis, in his Encyclical Letter *Laudato Si'* (Praise Be to You) in 2015 – a major doctrinal pronouncement for the Catholic Church – linked both our environmental crisis and widespread global poverty directly to what he characterized as "technocratic capitalism," and he called for a re-orientation of our economic system toward serving the "common good."[39] Later on that year, when the Pope addressed the U.S. Congress on this same theme, he referred to the "common good" no less than six times in his speech.

It may come as a surprise to learn that Adam Smith had similar views. Smith has often been stigmatized for the invisible hand model that he advanced in *The Wealth of Nations*. But this is literally taken out of context. In fact, Smith's moral foundation was the Stoic philosophy of world citizenship, the good of the community as a whole, and the Christian teaching of the Golden Rule. In his lesser-known early work, *The Theory of Moral Sentiments* (1759), Smith wrote that we should "love our neighbour as we love ourselves."[40] Moreover, according to Smith, virtue consists of exercising "self-command" over our baser impulses and having sympathy toward others.[41] Indeed, self-restraint is essential in a civilized society.[42] Furthermore, Smith believed (perhaps naively) that the invisible hand mechanism would benefit society as a whole because the rich would not consume a much greater proportion of the necessities of life; their share would only be of better quality.[43] In short, Smith was not promoting what has actually occurred, a system in which the rich often get richer at the expense of the poor – and at the expense of the global environment.

Shareholder Capitalism

The modern economic system commonly referred to these days as "shareholder capitalism" is therefore fundamentally at odds even with Smith's own moral values. It is deeply unfair in that it elevates the interests of business owners and shareholders over all the other interests that might have a stake in the success of a business firm, including sometimes the interests of society as a whole. Shareholder capitalism provides a license to be exploitative. In game theory terminology, it legitimizes a competitive zero-sum relationship rather than a win-win relationship. As a result, corporate business interests and practices have, in all too many cases, become a major contributor to the problems of global poverty and ecological destruction, rather than the solution.

Democratic governments and legal systems can and do play a role in counterbalancing this ideology in various ways, but far more must be done to re-orient our economic values and practices. To put it bluntly, the private sector must be subordinated to the common good. For some critics of capitalism, the answer is socialism. But, as I argue at length in *The Fair Society*, socialism in its more radical forms is also deeply unfair.[44] In any case, socialism is a political non-starter in today's world.

The alternative, I believe, is a middle-ground between capitalism and socialism. We must move toward a reformed and refocused industrial economy based on the concept of "stakeholder capitalism."[45] Although this is not a new idea, it must be pursued more aggressively as one of the keys to dealing with our global survival challenge.[46]

The concept of a stakeholder refers to anyone who has a material interest in a given business organization – in other words, when there is an economic *relationship* that entails costs and benefits for each of the participants. As a practical matter, it means that the actions of a business firm are likely to have an impact on the stakeholder's interests – for better or worse. For any large corporation in a modern complex society, the list of potential stakeholders is likely to be very long. It might include management personnel, various categories of workers, numerous subcontractors, many different suppliers (including transportation, energy, communications and Internet services), customers, local communities, multiple government entities, private financiers, and, of course, the shareholders. Indeed, even the interests of posterity might well be involved.

Stakeholder Capitalism

Accordingly, in the stakeholder capitalism model there must be structural arrangements (either formal or informal) that empower and advance the interests of all the various stakeholders, as appropriate. This might encompass such specific measures as worker and community representatives on the board of directors (as some countries already do), Fair Trade policies for suppliers, mechanisms for responding more effectively to customer feedback and complaints, a cooperative ongoing dialogue with suppliers, a mutualistic relationship with labor unions, and (especially in America), a greater willingness to accept legitimate government policies, regulations, and oversight when the public interest is at stake. (For instance, it would preclude sending profits offshore to avoid taxes, or paying poverty wages, or paying executive compensation at levels that amount to legalized looting, or, for that matter, fighting new air and water pollution regulations through the courts.)

The ideal outcome for the stakeholder model, as I and other supporters envision it, is to enhance a business firm's performance and its value by aligning and harmonizing the interests of the various stakeholders, rather than simply creating obstacles and roadblocks that might harm the company and reduce its value. This is far easier said than done, of course, but the principles of compromise and mutual accommodation, where needed, can achieve a great deal.

Indeed, a formal model (and analysis) developed by economist Franklin Allen and his colleagues in 2009 showed that such an alignment of interests is attainable – depending on the circumstances – and can lead to higher overall efficiency and value for a firm.[47] Their model is supported by a number of concrete examples, most notably in Germany, Japan, Austria, Luxembourg, and in some Nordic countries.

More recently, a new study in the U.S. has shown that American companies with narrower pay gaps between the CEO and the workers tend to perform better.[48] By the same token, there are also many thriving non-profit business firms and organized cooperatives these days, as well as a relatively new category of so-called B-Corporations that are committed to values that are compatible with the stakeholder model.

In order for stakeholder capitalism to thrive, however, there must also be a favorable business environment where "private equity firms" (corporate raiders) cannot prey on any company that does not maximize shareholder value, and where ruthless competitors cannot gain an unfair advantage. When all of the competitors in a particular marketplace must adhere to the stakeholder capitalism model, then no company is seriously disadvantaged by being fair to its various stakeholders and responsible toward society as a whole.[49] There is, in effect, a level playing field – as the saying goes. But in the context of our existing global economy, where corporate predators and sometimes unprincipled and exploitative competitors may act with complete disregard for the stakeholders or society as a whole, the playing field is often steeply tilted.

The Public Trust

Governments must therefore also play a greater role in shaping corporate behavior. This points us toward the third major change that I believe will be necessary for the next major transition in evolution, and here an ancient legal principle may be of some help.

It happens that the idea of a collective (societal) responsibility for the common good has a sturdy foundation in the concept of the public trust. This idea can be traced back to a category of Roman laws – *Jus publicum*, or public law – which (among other things) pertained to resources that were "by the law of nature" viewed as the common property of all humankind, including the air, water, the seas, and sea shores (according to the Institutes of Justinian).[50]

In the Medieval period, the idea of common ownership also came to be associated with such things as public thoroughfares and common pastures for grazing domestic animals. The principle that government has a responsibility and a role in protecting the commons is also embedded in English and American common law.

In modern times, the public trust doctrine has had many practical applications in various countries. In the U.S., the Federal government and number of states have used it to protect natural resources. The state of Washington, for instance, has mandated that all the fresh waters in the state are owned by the state as a common resource. Conditional "water rights" permits are required in order to use water for any large commercial purpose.[51]

There have also been many legislative applications of the public trust doctrine over the years. Important examples in the U.S. include the landmark National Environmental Policy Act (NEPA) in 1970, as well as the many federal laws over the years that have established some 59 national parks encompassing more than 51 million acres.

A Legal Tool

The public trust doctrine is also being used these days as a legal tool for fighting climate change and other environmental policy issues. For instance, in a bellwether case in 2013, the Pennsylvania Supreme Court found elements of that state's hydraulic fracturing legislation to be unconstitutional and in violation of the public trust. Currently pending is a lawsuit filed on behalf of 21 youths against the Federal government for violating their constitutional right to a healthy climate by supporting the production of fossil fuels and greenhouse gas emissions.[52] Regardless of the outcome of this case, it's highly significant that a Federal court has recognized its legitimacy (or "standing").

Another important application of the doctrine can be found in the so-called sovereign wealth funds, with Norway's very large fund as a premier example. These publicly managed funds are authorized to hold and invest discretionary state revenues, such as royalties from the sale of crude oil, in ways that are intended to benefit the common good.

"Legal Bedrock"

However, there is a deeper and broader interpretation of the public trust, championed by a number of legal scholars and some courts, that provides an opportunity for expanding its scope and application. The basic claim is that the public trust is a fundamental attribute of sovereignty in a democratic society – a "constitutive principle." It involves an inherent power to serve the public interest, and it has supremacy over contrary laws or individual property rights. As the University of Oregon law professor and public trust specialist Mary Christina Wood observes in her 2014 book, *Nature's Trust,* "characterizing the trust as an attribute of sovereignty bores down to legal bedrock."[53] In this interpretation, the public trust power and the ability to act in the public interest does not need to be backed by specific constitutional language or statutes. It no more needs to be spelled out than the police power, which is assumed to be a necessary element of sovereignty.

The concept of the common good is of similar character; it could be viewed as a basic responsibility of democratic governments. Professor Wood argues that, when government derives its power from the people, it necessarily imposes a fiduciary duty on the government to act as a trustee for the people.[54] Australian justice Paul Finn refers to it as the "inexorable logic of popular sovereignty."[55] Even the patron saint of private property rights, John Locke, observed that the "Fundamental, Sacred, and unalterable Law of Self-Preservation" forms the very basis of society and creates a responsibility for government to protect this right.[56]

A number of legal scholars also contend that this obligation should not be limited to the current generation. In Professor Wood's words: "The core purpose of the public trust lies in protecting the citizens' unyielding interest in their own survival (and that of their children)."[57] Similarly, Peter Brown in *Restoring the Public Trust,* asserts that "the trustees' fundamental duty is to preserve humanity."[58] And professor John Davidson points out that the core concern of America's founding fathers was the welfare of "posterity." Their intention when they wrote the Constitution was to create a social contract for the long term.[59]

It is, therefore, both logical and appropriate to conclude that the public trust encompasses whatever is required to sustain and advance the collective survival enterprise. All governments have a fiduciary responsibility to undergird and support the right to life and its indispensable corollary, a basic needs guarantee. Equally important, governments must impose a restraining and guiding influence on the private sector for the common good, or public interest, including the interests of posterity. To be sure, this is a hugely difficult task, amply confirmed by the disgraceful history of corrupt, captive, and self-serving governments over the past 10,000 years. But, to repeat Walter Lippmann's prescient long-ago warning, "Never before…have the stakes been so high."[60]

Finding a Way Forward

To summarize then, the way forward will depend on (1) a shift in our social values toward the Fair Society model, (2) major changes in our economic system toward stakeholder capitalism, and (3) governments that are empowered (and constrained) to act for the common good on behalf of the public trust.

It should be stressed that this vision of a Fair Society is emphatically *not* an unattainable ideal. There are some real-world examples. What has been called the Nordic Model – including especially Norway and some other Scandinavian countries – encompasses full employment at decent wages, a relatively flat distribution of income, a full array of supportive social services, extensive investment in infrastructure, excellent free education and health care, a generous retirement system, high social trust, a strong commitment to democracy, and a government that is sensitive to the common good, not to mention having a competitive capitalist economy with high productivity and deep respect for the environment. To top it off, Norway's sovereign wealth fund currently totals more than $1 trillion, a huge nest egg for such a small country. (Some apologists for American-style capitalism are dismissive about Norway, viewing it is an exception because it has the advantage of all those North Sea oil profits. Yes but, America was endowed with vastly greater oil deposits, which we have been exploiting for more than 100 years. So where is our sovereign wealth fund?) [61]

Beyond the few stellar examples like Norway, the challenge of realizing a global Fair Society is a daunting task, to say the least. When Albert Einstein was once asked why we were smart enough to produce atomic energy but couldn't contain nuclear weapons and the arms race, he answered: "It is because politics is more difficult than physics."[62]

A Species at Risk

Despite all of our past successes across countless generations of predecessors, it is evident that our species is at serious risk. In an increasing number of societies these days, the social contract is eroding or even breaking down. Especially alarming is the recent rise in many countries of xenophobia, extreme right-wing nationalism, and authoritarianism. America's turn to Donald Trump as President was a national tragedy and a symptom of its deep malaise. And the British decision to leave the European Union, along with the rise of virulent anti-immigration forces in Europe, is an ominous development.

Underlying these tectonic shifts, I believe, there is a deep economic insecurity – with many societies under stress and an underlying fear that there will not be enough to go around. As the old saying goes, when the pie gets smaller, the table manners change. Instead of responding positively to our ecological, economic and political challenges, there is the very real danger that we will seek to recapture a (presumably) safer, more comfortable world by retreating into an idealized tribal past and by building walls. But this will not work. It's a path that will dead end in division and lethal conflict.

Meanwhile, the ecological underpinnings of the survival enterprise will continue to be undermined. The twin scourges of widespread global poverty (and the social conflict that this engenders) and a mindlessly destructive technological system (or worse a system controlled by people who are in denial about what they are doing), coupled with the growing negative impact of climate warming, will ultimately force us to make radical changes – one way or the other. To borrow a warning from Lionel Shriver's dark new novel, "complex systems collapse catastrophically."[63] Jared Diamond's important book, *Collapse: How Societies Choose to Fail or Succeed*, vividly documents the many historical examples of bad social choices that have led to collapses in human societies. And, as Diamond warns, "Globalization makes it impossible for modern societies to collapse in isolation."[64] Ours would not be the first species to become the victims of our own success. Call it "niche destruction."

There are many ideas, proposals, and initiatives these days for how to solve our social and ecological crises – how to transition to a more "sustainable" global society (to use the current buzzword). Some ideas are very promising. Others are half-hearted palliatives or, worse, placebos. In any case, we must act with the kind of urgency that would make historic undertakings like the Manhattan Project in World War Two, or the Apollo space program in the 1960s, look like warm up exercises. It will require a major shift of values and concerted action on a global scale – a radical change in the political status quo.

Starting Yesterday

For instance, renewable, non-polluting energy production – mainly wind and solar power – currently stands at about 8% of the global total (excluding hydroelectric power).[65] To avoid calamitous global warming within the next two generations, it is estimated that we will need roughly a 40-fold increase in wind and solar energy capacity, starting yesterday. (Fortunately, the cost of renewable energy has now become highly competitive, and even the cost of energy storage equipment is rapidly coming down.)[66]

The Paris climate conference in December 2015 represented a hopeful start toward addressing the problem, but the measures that were agreed to there still fall far short of avoiding a long-term disaster. The pledges made in Paris by some 187 countries would still result in global warming of around 3^0 Celsius above the pre-industrial level, or twice the 1.5^0C. cap that the conference attendees themselves set as the necessary goal. The Paris agreement was also voluntary, and it remains to be seen if the many individual commitments for change will be honored.[67] Worse yet, it now appears that the climate cause has lost the vital impetus of American leadership. Meanwhile, the global warming trend continues relentlessly, with 2014, 2015, and 2016 being the three hottest years on record. As predicted, they were accompanied by many extreme local weather events.

The Challenge of Artificial Intelligence

It is one of the great ironies of our age that our technological progress – for all its many benefits – is becoming an increasingly serious threat to our future in still another way with the development of artificial intelligence (AI) and robotics. As the venture capitalist and AI expert Kai-Fu Lee warned in a recent *New York Times* op ed article, we are approaching a technological "singularity" – a point where many tens of millions of workers – from bank tellers to taxi drivers, construction workers, and many others – will be replaced by smart machines and will be unable to find alternative work.[68] The need for a radically new social contract will become an imperative without precedent historically.

The collective survival enterprise in humankind is also an increasingly interdependent global system, so we must mobilize ourselves to create a new, globe-spanning biosocial contract. Former President Obama, in his Hiroshima speech, characterized it as a "moral revolution." At the heart of the next major transition in evolution, there must be a unifying global vision, and unifying social values. And, like all the previous major transitions, this one will require a new level of organization and cooperation – and governance – that can foster new forms of synergy and Synergistic Selection on a global scale.

Given the deep tribal roots and polarizing national loyalties that divide our emerging global society, coupled with the profound religious, cultural, economic, linguistic, and political divisions that serve to reinforce these tendencies (not to mention the many direct conflicts of interest), the idea of a major transition to global governance may seem totally unrealistic. In a timely article about our political dilemma in the *Proceedings of the National Academy of Sciences*, science writer Stephen Battersby asks: "Lacking any higher authority to rein in the selfishness of nations, are we doomed?"[69] He sees no silver bullet.

The Tragedy of the Commons

The overarching problem we face was framed for us in an inspired metaphor that Garrett Hardin used in his classic 1968 article on "The Tragedy of the Commons."[70] Whenever there is a limited resource that everyone is free to exploit – like the public grazing pastures in Medieval times, or a global Commons like our oceans and atmosphere – eventually it will be over-exploited and destroyed. As Hardin memorably warned: "Freedom in a Commons brings ruin to all." Limits must ultimately be set.

For Hardin, the only viable solution was "mutual coercion mutually agreed upon" – democratic and consensual self-regulation. Hardin was most concerned about population growth, and he foresaw the need for top-down government intervention (as China has in fact done, albeit without mutual consent). Decades of subsequent research and theoretical work on the Commons issue by the Nobel Prize winning political scientist Elinor Ostrom and her many students and colleagues has shown that there are various alternative ways of regulating "common pool resources" – as Ostrom called them. In the end, she concluded that both top-down and bottom-up approaches to regulation can be effective, depending on the nature of the problem and how they are executed. But, in any case, governance is essential – at every level. To deal with our environmental crisis, Ostrom recommended a "polycentric approach" that engages all levels of local, regional and national stakeholders, including shared responsibility and multiple strategies.[71]

An "Aha" Moment

To achieve these political changes, however, there will need to be a transformative change of perceptions and understanding – perhaps an "aha" moment when the global threat is universally recognized for what it is and the imperative for collective action on a global scale becomes obvious and compelling in all parts of the global community. For those of us who are old enough to remember, it would be the psychological equivalent of "Pearl Harbor" – the surprise air attack on America's Pacific fleet in World War Two that galvanized a divided nation.

Going forward, there must also be inspired and skillful leadership and a broad mobilization of public support and action in every country to fight what must become a global "war" on environmental destruction, unemployment, and poverty. Drastic collective action on climate change and the environment coupled with a shift to stakeholder capitalism and a global basic needs guarantee is the only (consensual) way forward: "mutual coercion mutually agreed upon." Such a consensus seems at present far out of reach.

A New Global Order

Ultimately, this change implies a new level of collective self-government and a new political order. The necessary institutional structures have yet to be invented, but there are precedents, including the Constitutional Convention in 1787 that led to the creation of the United States, the San Francisco conference in 1945 that established the United Nations, and, more recently, the creation of the European Union.[72] Perhaps a new global constitution – a common legal system and a global structure for collective action – can be erected on the foundation provided by the U.N., and the International Court of Justice. But this must be accompanied by a global effort to achieve a Fair Society, for this is the only way to mitigate, if not dissolve the deep political divisions that stand in the way. Is this an unrealistic goal? It can only be called unrealistic if you fail to consider the likely alternative. Indeed, it is no more unrealistic than all the other major transitions in evolution, including the rise of humankind.[73]

The idea of world government is, of course, hardly new. It's a dream that can be traced back at least to Bronze Age Egypt and the ancient Chinese Emperors. In the modern era, it has been espoused by a great many prominent figures, from Dante to Immanuel Kant to Winston Churchill. It could be said that the League of Nations and the United Nations were baby steps in this direction, but what was once an aspiration has now become an imperative. To paraphrase the great journalist and peace advocate Norman Cousins, our security will ultimately be found only through the mutual control of force, not the use of force – no small challenge.[74] David Sloan Wilson concludes his new book *Does Altruism Exist?* with this exhortation: "If we want the world to become a better place, we must choose policies with the welfare of the whole world in mind…We must become planetary altruists."[75]

A Global Superorganism

A major theme in this book has been the creative role that living organisms themselves have played in shaping the course of evolution, culminating in the rise of the Self-Made Man. Now the Self-Made Man must take the initiative and evolve into a global superorganism. Otherwise, we face the possibility of what could perhaps be termed the Anthropocene Implosion.[76] There will be growing political conflict and social chaos, horrific violence and human suffering, and wanton self-annihilation on a global scale – not to mention the destruction of the biosphere (and our life-support system) as we know it. In a 2012 article in *The Proceedings of the Royal Society* co-authored by Paul Ehrlich and his wife (and long-time colleague) Anne Ehrlich, the authors ask: "Can a collapse of global civilization be avoided?" It is possible, they conclude, but they are not optimistic.[77]

An ominous foretaste of this dark future scenario is the current turmoil in the Middle East – ranging from the (mostly) disastrous Arab Spring to the Syrian civil war, the rise of ISIS and the flood of refugees – all of which may in fact have been triggered by severe droughts and steep spikes in global food prices, according to a new in-depth analysis.[78] (It's also important to remember that this existential threat is greatly amplified by the proliferation of nuclear and biological weapons, and by the global reach of long-range missiles – not to mention the dark menace of ruthless terrorists.) As Edward Wilson put it in a recent interview, "We are a dysfunctional species," with "Paleolithic emotions, Medieval institutions and god-like technologies…That's a dangerous mix."[79]

Synergy is the Way Forward

The next major transition in evolution must span the entire globe and must subordinate the entire human species to the pursuit of the "common good" – which, again, can be defined in biological terms as sustaining and enhancing our interdependent "collective survival enterprise." In the final reckoning, if our species fails to meet this great survival challenge, we will squander our evolutionary inheritance and betray what untold generations of our ancestors struggled to achieve over millions of years.[80] To paraphrase the American founding father, Benjamin Franklin, we must all survive together or we will go extinct separately.[81]

Our generation confronts an inescapable collective choice. If we can achieve global governance and a Fair Society for our species as a whole and, in the bargain, ensure the future of life on Earth as we know it, this would indeed be another major transition in evolution and, equally significant, a transcendent example of Synergistic Selection. One of the great take-home lessons from the epic of evolution is that cooperation produces synergy, and synergy is the way forward. The arc of evolution bends toward synergy.

---

[1] Quoted in Stringer 2012, p. 263.

[2] See especially Wills 1998; also, Stock 2002; Hawks *et al.* 2007; Cochran and Harpending 2009; Lieberman 2013; Thomas 2015. See also the op-ed article in the *New*

York Times by Menno Schilthuizen, Evolution Is Happening Faster Than We Thought. http://www.nytimes.com/2016/07/24/opinion/sunday/evolution-is-happening-faster-than-we-thought.html?emc=eta1 (accessed 24 July 2016). If further evidence of rapid evolution is needed, consider the diversity of domesticated dogs.

[3] Hawks *et al.* 2007; also, Stringer 2012, pp. 265-269; Lieberman 2013, p. 205.

[4] The evidence for such micro-evolutionary change as an ongoing process in human evolution was reviewed by Wills 1998; see also Wolpoff 1999a; Lieberman 2013. New DNA evidence for genetic changes associated with the agricultural revolution was reported by Mathieson *et al.* 2015. See also "Not what they were: Researchers can now watch human evolution unfold." *The Economist,* May 14, 2016, pp. 71-72.

[5] Laland *et al.* 2010. On the other hand, we also suffer from an array of what Daniel Lieberman (2013, pp. 168-174, 202) calls "mismatch diseases." (See the outtake at my website.)

[6] https://en.wikipedia.org/wiki/Anthropocene (last modified 27 January 2016). (See the outtake on this at my website.) Some ecologists have characterized humankind as "the world's greatest evolutionary force." See Hendry *et al.* 2016. For an analysis of the ecological consequences, see Boivin *et al.* 2016.

[7] There are a great many different kinds of negative synergy (or "dysergy") in the natural world – cooperative effects that are deleterious to one or more participants, or to various "bystanders". Whether synergy can be called "positive" or "negative" depends, quite simply, on the value that is assigned to it. For instance, cooperative hunting might be considered very beneficial by a group of predators, though the outcomes would be viewed as negative synergy by their prey. Parasitism provides innumerable examples of such differences in perspective. (Much more on negative synergy can be found in Corning, 2003.)

[8] This estimate was made by the United Nations' Food and Agriculture Organization. See https://www.scientificamerican.com/article/only-60-years-of-farming-left-if-soil-degradation-continues/ (accessed 2 July 2017).

[9] Among others, see Wills 1998; Stock 2002, 2007; Cochran and Harpending 2009; Lieberman 2013; Thomas 2015. See also the op-ed article in the *New York Times* by Menno Schilthuizen, Evolution Is Happening Faster Than We Thought. http://www.nytimes.com/2016/07/24/opinion/sunday/evolution-is-happening-faster-than-we-thought.html?emc=eta1 (accessed 24 July 2016). If further evidence of rapid evolution is needed, consider domesticated dogs. On the Indian drought, see http://www.bbc.com/news/world-asia-india-36089377 (accessed 5 May 2016).

[10] The current target for containing global warming falls far short. (For more details, see the outtake at my website.)

[11] The original version was coined by economist Herbert Stein: "If something cannot go on forever, it will stop." See http://www.goodreads.com/quotes/467811-if-something-cannot-go-on-forever-it-will-stop (accessed 16 May 2016). A similar idea was later expressed by economist Kenneth Boulding: "Anyone who believes that exponential

growth can go on forever in a finite world is either a madman or an economist." http://www.goodreads.com/quotes/627148-anyone-who-believes-that-exponential-growth-can-go-on-forever (accessed 16 May 2016).

[12] Ehrlich and Harte 2015. They cite at least eight major "top of our list" policy changes, ranging from a global carbon tax to drastically reducing the use of pesticides and other chemicals in agriculture. They conclude "We find it hard to be optimistic."

[13] Cited in Ehrlich and Ehrlich 2012.

[14] http://www.nytimes.com/2016/05/28/world/asia/text-of-president-obamas-speech-in-hiroshima-japan.html?_r=0 (accessed 28 May 2016).

[15] Kiers and West 2015. (A lengthy quote from the authors' paper can be found as an outtake at my website.)

[16] Richerson and Boyd 1999. See also Boehm 1997.

[17] These fourteen categories are detailed and documented in Corning 2011.

[18] It should be noted that this list is convergent with U.N.'s Human Development Index. (For more on this, see the outtake at my website.)

[19] Equality has been a socialist and liberal/progressive ideal ever since the Enlightenment. The shortcomings of egalitarian socialism are discussed in some detail in my 2011 book. (For a brief discussion, see also the outtake at my website.)

[20] Frohlich and Oppenheimer 1992.

[21] See https://www.washingtonpost.com/news/wonk/wp/2016/04/25/bernie-sanders-is-profoundly-changing-how-millennials-think-about-politics-poll-shows/?tid=a_inl (accessed 10 May 2016).

[22] See https://en.wikipedia.org/wiki/Guaranteed_minimum_income (last modified 25 March 2017). Among others, supporters of the idea have included Thomas Paine, Henry George, Milton Friedman, Friedrich Hayek, Martin Luther King, Daniel Patrick Moynihan, and a manifesto signed by 1200 economists in 1968, led by James Tobin, Paul Samuelson, and John Kenneth Galbraith.

[23] Locke 1970/1690.

[24] https://www.marxists.org/reference/archive/smith-dam/works/moral/part02/part2b.htm (accessed 28 January 2016).

[25] http://www.utm.edu/research/iep/b/bentham.htm (accessed 28 January 2016).

[26] Societies have long qualified this right, or hedged it in various ways, including the killing in war (or Jihad), capital punishment, euthanasia, self-defense, etc.

[27] Although it's still a third rail politically, there is a long-term trade off that may be necessary – namely, global population control. (For more on this controversial issue, see the outtake at my website.)

[28] Plato. 1946/380 B.C.

[29] A much longer discussion of this issue can be found in *The Fair Society*, Chapter 6. For critiques of the neo-classical model, see Beinhocker 2006; Bowles 2004; Gowdy *et al.* 2013.

[30] The reality is, of course more complex. Wealth accumulation is a proven method for stimulating further innovation and entrepreneurship, but it also leads to conspicuous personal consumption, idled wealth, and other wasteful outcomes. Moreover, there is another alternative. Public taxes on wealth, if properly invested, can accomplish the same end.

[31] Smith 1964/1776, IV, 2, 9. Modern economists often become lyrical about "the superiority of "self-interest" over altruism in economic life and the virtues of competition and the "profit motive," while overlooking the fact that Smith's rendering of the invisible hand was quite contingent. (For more on Smith, see the outtake at my website.)

[32] Mainstream economists might argue that utopian capitalism no longer reflects the orthodoxy that was once predominant. (For more, see the outtake at my website.)

[33] Bowles 2004, p. 208.

[34] Gowdy, ed. 1998, pp. xvi-xvii. Of course, many liberal economists have challenged the neo-classical model. (For more, see the outtake at my website.)

[35] Corning 2011.

[36] Piketty 2014.

[37] Korten 2015/1995.

[38] See the insightful edited volume on this subject, *Moral Markets: The Critical role of Values in the Economy*, edited by Paul Zak 2008; also, Geoffrey Hodgson's *Conceptualizing Capitalism* (2015); also, Beinhocker 2006.

[39] Pope Francis 2015. Papal Encyclical *Laudato Si'* ("praise be to you") p. 54. http://w2.vatican.va/content/francesco/en/encyclicals/documents/papa-francesco_20150524_enciclica-laudato-si.html (accessed 16 July 2015).

[40] Smith 1976/1759, Vol. I.i.5.5. See also note 4; also, Ritter 1954.

[41] *Ibid.*, Vol. II.iii.34.

[42] *Ibid.*, Vol. VI.iii.2.

[43] *Ibid.*, Vol. IV.i.10.

[44] In a nutshell, socialism in theory is cavalier about the principle of *equity* – differential rewards (or punishments) for merit and is vague about the principle of *reciprocity.* See Corning 2011.

[45] An overview can be found in Kelly *et al.* eds. 1997; also, Ackerman and Alstott 1999. (For more on this concept, see the outtake at my website.)

[46] In fact, the spirit of stakeholder capitalism long predates the term. (For details, see the outtake at my website.)

[47] Franklin Allen, *et al.*, "Stakeholder Capitalism, Corporate Governance, and Firm Value." (September 2009). http://fic.wharton.upenn.edu/fic/papers/09/0928.pdf.

[48] See http://www.nytimes.com/2016/05/11/opinion/pressure-to-close-the-pay-gap.html?emc=edit_th_20160511&nl=todaysheadlines&nlid=27277596&_r=0 (accessed 11 May 2016).

[49] Government mandates, such as minimum wage legislation and regulations governing working hours, sick leave, etc., or the legal protections afforded by B-corporation status,

also have the effect of leveling the playing field. Even Walmart, a firm that is notorious for paying poverty wages, must pay higher wages in jurisdictions that require it.

[50] Wood 2014. See also https://en.wikipedia.org/wiki/Public_trust_doctrine (last modified 11 September 2015).

[51] Both federal and state courts in the U.S. have also recognized the public trust in various rulings. (For more details, see the outtake at my website.)

[52] See https://www.washingtonpost.com/news/energy-environment/wp/2017/07/05/trashed/

[53] Wood 2014, p. 132.

[54] *Ibid.*, p. 128.

[55] Quoted in Wood, *op cit.*, p. 128.

[56] Locke, 1970/1690.

[57] Wood, *op. cit.*, p. 126.

[58] Brown 1994, p. 78.

[59] Cited in Wood, *op. cit.*, pp. 129-130.

[60] Quoted in Brandon 1969.

[61] Norway has also used its profits to make many improvements in its infrastructure and public services, and it is now winding down the oil sector of its economy.

[62] http://todayinsci.com/E/Einstein_Albert/EinsteinAlbert-PoliticsQuote500px.htm (accessed 28 January, 2016).

[63] Shriver 2016. For documentation, see Tainter 1988; Diamond 2005; Corning 2005 (Chapter 7). For a new mathematical model that reveals some of the underlying network dynamics, see Yu *et al.* 2016.

[64] Diamond 2005, p. 23.

[65] See the analysis in *The Economist* (Volume 416, Number 8949, 1 August 2015, pp. 12-13); also *The Economist* (Volume 416, Number 8949, 15 July 2017, p. 12); also http://www.nytimes.com/2016/04/04/opinion/a-renewable-energy-boom.html?emc=edit_th_20160404&nl=todayheadlines&nlid=27277596&_r=0 (accessed 10 May 2016). (For more details, see the outtake at my website).

[66] http://thinkprogress.org/climate/2016/05/09/3775606/used-second-life-electric-car batteries/?utm_source=newsletter&utm_medium=email&utm_campaign=tptop3&utm_term=1&utm_content=53&elqTrackId=e7585cedc9b44b26950cb980701cc136&elq=f7aba5fa560e44ee861afaec283cfb5c&elqaid=30054&elqat=1&elqCampaignId=5581 (accessed 9 May 2016).

[67] *The Economist*, December 19, 2015–January 1, 2016, p. 89.

[68] https://www.nytimes.com/2017/06/24/opinion/sunday/artificial-intelligence-economic-inequality.html

[69] Battersby 2017.

[70] Hardin 1968.

[71] Ostrom 2009. (See the outtake on Ostrom's Eight Principles for common resource management at my website.)

[72] Of course, each of these historic acts of political creativity was built on a foundation of precursor institutions – namely, the Articles of Confederation and the League of Nations. A similar foundation may exist today in the United Nations.

[73] In an important article on how groups can evolve and prevail, David Sloan Wilson and Edward Wilson (2007) put it this way: "Selfishness beats altruism within groups. Altruistic groups beat selfish groups. Everything else is commentary." A full-length elaboration on this thesis can be found in D.S. Wilson 2015.

[74] The problem of achieving global government has long been debated by political scientists, and others. (See the outtake on this at my website.)

[75] D.S. Wilson 2015, p. 149.

[76] Jared Diamond (2005) asks why many societies historically have collapsed while others have endured. (For more, see the outtake at my website.)

[77] Ehrlich and Ehrlich 2012.

[78] Lagi *et al.* 2015. (For details about this study, see the outtake at my website.) Two other recent studies reinforce this climate-conflict nexus. See Schleussner *et al.* 2016; von Uexkull *et al.* 2016.

[79] "E.O. Wilson, Of Ants and Men," premiered on PBS 30 September 2015. http://www.pbs.org/program/eo-wilson/ (accessed 28 January 2016).

[80] It should also be noted that some contemporary theorists portray a global cooperative community as an inexorable trend. I disagree. (See the outtake on this at my website.)

[81] Franklin's famous remark, spoken just before signing the American Declaration of Independence in 1776, was "We must, indeed, all hang together or, most assuredly, we shall all hang separately." As traitors! See https://www.brainyquote.com/quotes/quotes/b/benjaminfr151597.html (assessed 13 August 2017).

References

Abbot, Patrick, *et al.* 2011. Inclusive Fitness Theory and Eusociality ("Brief Communications Arising"). *Nature* **471**: E1-E10.

Ackerman, Bruce, and Anne Alstott. 1999. *The Stakeholder Society.* New Haven, CT: Yale University Press.

Aiello, Leslie C., and Robin I.M. Dunbar. 1993. Neocortex Size, Group Size, and the Evolution of Language. *Current Anthropology* **34**: 184-192.

Aiello, Leslie C., and Peter Wheeler. 1995. The Expensive-Tissue Hypothesis. *Current Anthropology* **36**: 199-221.

Aiello, Leslie C., and Cathy Key. 2002. The Energetic Consequences of Being a Homo erectus Female. *American Journal of Human Biology* **14**(5): 551-565.

Alemseged, Zeresenay, *et al.* 2006. A Juvenile Early Hominin Skeleton from Dikika, Ethiopia. *Nature* **443**: 296-301.

Alexander, Richard D. 1987. *The Biology of Moral Systems.* New York: Aldine de Gruyter.

Allee, Warder C. 1931. *Animal Aggregations: A Study in General Sociology.* Chicago, IL: University of Chicago Press.

Allen, Benjamin, Martin A. Nowak, and Edward O. Wilson. 2013. Limitations of Inclusive Fitness. *Proceedings of the National Academy of Sciences* **110**: 20135-20139.

Allen, Judy R. M., *et al.* 1999. Rapid Environmental Changes in Southern Europe During the Last Glacial Period. *Nature* **400**: 740-743.

Allen, Mark, Robert Lawrence Bettinger, Brian F. Codding, Terry L Jones, and Al W. Schwitalla. 2016. Resource Scarcity Drives Lethal Aggression Among Prehistoric Hunter-Gatherers in Central California. *Proceedings of the National Academy of Sciences* **113**: 12120-12125.

Allen, Pamela. 1983. *Who Sank the Boat?* New York: Coward-McCann.

Alley, Richard B. 2000. Ice-core Evidence of Abrupt Climate Changes. *Proceedings of the National Academy of Sciences* **97**: 1331-1334.

Allison, David G., Peter Gilbert, Hilary M. Lappin-Scott, and Michael Wilson. 2001. *Community Structure and Cooperation in Biofilms.* London: Pergamon Press.

Ambrose, Stanley. 2001. Paleolithic Technology and Human Evolution. *Science* **291**: 1748-1753.

Anderson, Carl, and Nigel R. Franks. 2001. Teams in Animals Societies. *Behavioural Ecology* **12**(5): 534-540.

Anderson, Connie M. 1986. Predation and Primate Evolution. *Primates* **27**: 15-39.

Apicella, Coren L., Frank W. Marlowe, James H. Fowler, and Nicholas A. Christakis. 2012. Social Networks and Cooperation in Hunter-Gatherers. *Nature* **481**: 497-501.

Aristotle. 1946/Ca. 350 B.C. *Politics* (trans. Ernest Barker). Oxford: Oxford University Press.

______ . 1961/Ca. 350 B.C. *The Metaphysics* (trans. H. Tredennick). Cambridge, MA: Harvard University Press.

Asfaw, Berhane, *et al.* 1999. *Australopithecus garhi:* A New Species of Early Hominid from Ethiopia. *Science* **284**: 629-635.

Avital, Eytan, and Eva Jablonka. 2000. *Animal Traditions: Behavioral Inheritance and Evolution.* Cambridge, UK: Cambridge University Press.

Bardi, Ugo. 2014. *Extracted: How the Quest for Mineral Wealth Is Plundering the Planet.* White River Junction, VT.: Chelsea Green Publishing.

Bar-Yosef, Ofer. 2002. The Upper Paleolithic Revolution. *Annual Review of Anthropology* **31**: 363-393.

Bateson, Patrick P.G. 1988. The Active Role of Behavior in Evolution. In *Evolutionary Processes and Metaphors,* ed. Mae-Wan Ho, and Sidney W. Fox, 191-207. New York: John Wiley & Sons.

______ . 2004. The Active Role of Behaviour in Evolution. *Biology and Philosophy* **19**: 283-298.

______ . 2005. The Return of the Whole Organism. *Journal of Bioscience* **30**: 31-39.

______ . 2013. Evolution, Epigenetics and Cooperation. *Journal of Bioscience* **38**(4): 1-10.

Bateson, Patrick P.G., and Robert Hinde, eds. 1976. *Growing Points in Ethology*. London: Cambridge University Press.

Bateson, Patrick P.G., and Peter Gluckman. 2011. *Plasticity, Robustness, Development, and Evolution.* Cambridge, UK: Cambridge University Press.

Battersby, Stephen. 2017. Can Humankind Escape the Tragedy of the Commons? *Proceedings of the National Academy of Sciences* **114**(1): 7-10.

Beck, Benjamin B. 1980. *Animal Tool Behavior.* New York: Garland Press.

Bedau, Mark A. 2012. A Functional Account of Degrees of Minimal Chemical Life. *Synthese* **185**: 73-88.

Beinhocker, Eric D. 2006. *The Origin of Wealth: Complexity and the Radical Remaking of Economics.* Boston: Harvard Business School Press.

Bell, Adrian V., Peter J. Richerson, and Richard McElreath. 2009. Culture Rather Than Genes Provides Greater Scope for the Evolution of Large-Scale Human Prosociality. *Proceedings of the National Academy of Sciences* **106**: 17671-17674.

Bell, Graham. 1985. Origin and Early Evolution of Germ Cells as Illustrated by the Volvocales. In *Origin and Evolution of Sex,* ed. Harlyn O. Halverson, and Alberto Monroy, 221-256. New York: Alan R. Liss.

Bellomo, Randy. 1994. Methods of Determining Early Hominid Behavioral Activities Associated with the Controlled Use of Fire at FxJi Main, Koobi Fora, Kenya. *Journal of Human Evolution* **27**: 173-195.

Bergquist, Patricia R. 1978. *Sponges.* Berkeley, CA: University of California Press.

Berna, Francesco, *et al.* 2012. Microstratigraphic Evidence of in Situ Fire in the Acheulean Strata of Wonderwerk Cave, Northern Cape Province, South Africa. *Proceedings of the National Academy of Sciences* **109**: E1215-E1220.

Bickerton, Derek. 2009. *Adam's Tongue: How Humans Made Language, How Language Made Humans.* New York: Hill and Wang.

Bickerton, Derek, and Eörs Szathmáry. 2011. Confrontational Scavenging as a Possible Source for Language and Cooperation. *BMC Evolutionary Biology* **11**: 261-268.

Binford, Lewis R. 1987. The Hunting Hypothesis: Archeological Methods, and the Past. *American Journal of Physical Anthropology* **30** (Suppl. S8): 1-9.

Bingham, Paul M. 2000. Human Evolution and Human History: A Complete Theory. *Evolutionary Anthropology* **9**: 248-257.

Binmore, Ken. 2005. *Natural Justice.* Cambridge, MA: MIT Press.

Blitz, David. 1992. *Emergent Evolution: Qualitative Novelty and the Levels of Reality.* Dordrecht, Netherlands: Kluwer Academic Publishers.

Blumenschine, Robert J. 1987. Characteristics of an Early Hominid Scavenging Niche. *Current Anthropology* **28**: 383-407.

Blumenschine, Robert J., John A. Cavallo, and Salvatore D. Capaldo. 1994. Competition for Carcasses and Early Hominid Behavioral Ecology: A Case Study and Conceptual Framework. *Journal of Human Evolution* **27**: 197-213.

Blumenschine, Robert J., Kari A. Prassack, C. David Kreger, and Michael C. Pante. 2007. Carnivore Tooth-marks, Microbial Bioerosion, and the Invalidation of Domínguez-Rodrigo and Barba's (2006) Test of Oldowan Hominin Scavenging Behavior. *Journal of Human Evolution* **53**: 420-426.

Boehm, Christopher. 1993. Egalitarian Behavior and Reverse Dominance Hierarchy. *Current Anthropology* **34**: 227-254.

______ . 1996. Emergency Decisions, Cultural-Selection Mechanics, and Group Selection. *Current Anthropology* **37**: 763-793.

______ . 1997. Impact of the Human Egalitarian Syndrome on Darwinian Selection Mechanics. *The American Naturalist* **150**: 100-121.

______ . 1999. *Hierarchy in the Forest: The Evolution of Egalitarian Behavior.* Cambridge, MA: Harvard University Press.

Boesch, Cristophe, and Michael Tomasello. 1998. Chimpanzees and Human Cultures. *Current Anthropology* **39**: 591-614.

Bogucki, Peter. 1999. *The Origins of Human Society.* Oxford: Blackwell.

Boivin, Nicole L. *et al.* 2016. Ecological Consequences of Human Niche Construction: Examining Long-Term Anthropogenic Shaping of Global Species Distributions. P*roceedings of the National Academy of Sciences* **113**: 6388-6396.

Bokma, Folmer, Valentijn van den Brink, and Tanja Stadler 2012. Unexpectedly Many Extinct Hominins. *Evolution* **66**: 2969-2974.

Bonabeau, Eric, Marco Dorigo, and Guy Theralulaz. 1999. *Swarm Intelligence: From Natural to Artificial Systems.* New York: Oxford University Press.

Bonner, John Tyler. 1988. *The Evolution of Complexity by Means of Natural Selection.* Princeton, NJ: Princeton University Press.

______. 1999. The Origins of Multicellularity. *Integrative Biology* **1**: 27-36.

______. 2003. On the Origin of Differentiation. *Journal of Bioscience* **28**: 523-528.

______. 2006. *Why Size Matters: From Bacteria to Blue Whales.* Princeton, N.J.: Princeton University Press.

Bowles, Samuel. 2004. *Microeconomics: Behavior, Institutions, and Evolution.* New York: Russell Sage Foundation/ Princeton University Press.

______. 2009. Did Warfare Among Ancestral Hunter-Gatherers Affect the Evolution of Human Social Behaviors? *Science* **324**: 1293-1298.

Bowles, Samuel, and Herbert Gintis. 2011. *A Cooperative Species: Human Reciprocity and Its Evolution.* Princeton, NJ: Princeton University Press.

Boyd, Robert, and Peter J. Richerson. 1985. *Culture and the Evolutionary Process.* Chicago: University of Chicago Press.

Boyd, Robert, and Peter J. Richerson. 2005. *The Origin and Evolution of Cultures.* Oxford: Oxford University Press.

Boyd, Robert, and Peter J. Richerson. 2009. Culture and the Evolution of Human Cooperation. *Philosophical Transactions of the Royal Society B* **364**: 3281-3288.

Boyd, Robert, Peter J. Richerson, and Joseph Henrich. 2011. The Cultural Niche: Why Social Learning in Essential for Human Adaptation. *Proceedings of the National Academy of Sciences* **108**: 10918-10925.

Boyd, Robert, Peter J. Richerson, and Joseph Henrich. 2013. Cultural Evolution of Technology: Facts and Theories. In *Cultural Evolution: Society, Technology, Language, and Religion,* ed. Peter J. Richerson, and Morten Christiansen, 119-142. Cambridge, MA: MIT Press.

Brain, Charles K. 1981. *The Hunters or the Hunted?* Chicago, IL: University of Chicago Press.

______. 1985. Interpreting Early Hominid Death Assemblies: The Use of Taphonomy since 1925. In *Hominid Evolution: Past, Present and Future: Proceedings of the Taung Diamond Jubilee International Symposium,* ed. Phillip V. Tobias, 41-46. New York: Alan R. Liss, Inc.

Bramble, Dennis M., and Daniel E. Lieberman. 2004. Endurance Running and the Evolution of *Homo. Nature* **432**: 345-352.

Brandon, Henry. 1969. A Talk with Walter Lippmann at 80, About This 'Minor Dark Age'. *New York Times Magazine* September 14, 1969, pp. 25 ff.

Breed, Michael D., and Janice Moore, eds. 2010. *Encyclopedia of Animal Behavior.* London: Academic Press.

Brooks, Allison, *et al.* 1995. Dating and Context of Three Middle Stone Age Sites with Bone Points in the Upper Semliki Valley, Zaire. *Science* **268**: 548.

Brown, Donald E. 1991. *Human Universals.* Philadelphia: Temple University Press.

Brown, Peter G. 1994. *Restoring the Public Trust: A Fresh Vision for Progressive Government in America.* Boston: Beacon Press.

Bunn, Henry T. 2001. Hunting, Power Scavenging and Butchery by Hadza Foragers and by Plio-Pleistocene Homo. In *Meat-eating and Human Evolution,* ed. Craig B. Stanford, and Henry T. Bunn, 199-217. New York: Oxford University Press.

Burkart, Judith M., Sarah B. Hrdy, and Carel P. van Schaik. 2009. Cooperative Breeding and Human Cognitive Evolution. *Evolutionary Anthropology* **18**: 175-186.

Burkart, Judith M. *et al.* 2014. The Evolutionary Origin of Human Hyper-Cooperation. *Nature Communications* **5**: 4747; doi: 10.1038/ncomms5747

Buss, Leo W. 1987. *The Evolution of Individuality.* Princeton, NJ: Princeton University Press.

Byrne, Richard W. 1995. *Thinking Apes: The Evolutionary Origins of Intelligence.* Oxford: Oxford University Press.

Byrne, Richard W., and Andrew Whiten, eds. 1988. *Machiavellian Intelligence: Social Expertise and the Evolution of Intellect in Monkeys, Apes and Humans.* Oxford: Clarendon Press.

Cairns-Smith, Alexander Graham. 1987/1982. *Genetic Takeover and the Mineral Origins of Life.* Cambridge, UK: Cambridge University Press.

______ . 2009/1968. An Approach to a Blueprint for a Primitive Organism. In *The Origin of Life: Towards a Theoretical Biology* (a reissue), ed. Conrad H. Waddington, Vol. 1, 57-66. Edinburgh: Edinburgh University Press.

Calcott, Brett. 2013a. Why How and Why Aren't Enough: More Problems with Mayr's Proximate-Ultimate Distinction. *Biology and Philosophy* **28**(5): 767-780.

______ . 2013b. Why the Proximate–Ultimate Distinction is Misleading, and Why It Matters for Understanding the Evolution of Cooperation. In *Cooperation and Its Evolution,* ed. Kim Sterelny, Richard Joyce, Brett Calcott, and Ben Fraser, 249-263. Cambridge, MA; MIT Press.

Calcott, Brett, and Kim Sterelny, eds. 2011. *The Major Transitions in Evolution Revisited.* Cambridge, MA: MIT Press.

Callahan, Michael P., *et al.* 2011. Carbonaceous Meteorites Contain a Wide Range of Extraterrestrial Nucleobases. *Proceedings of the National Academy of Sciences* **108**: 13995-13998.

Campbell, Bernard. 1985. *Human Evolution: An Introduction to Man's Adaptations* (3rd edn.). New York: Aldine.

Campbell, Donald T. 1974. Downward Causation in Hierarchically Organized Biological Systems. In *Studies in the Philosophy of Biology,* ed. Theodosius Dobzhansky and Francisco J. Ayala, 85-90. Berkeley, CA: University of California Press.

Campbell, John H. 1994. Organisms Create Evolution. In *Creative Evolution?!* ed. John H. Campbell, and J. William Schopf, 85-102. Boston: Jones & Bartlett.

Cannon, Herbert G. 1955. What Lamarck Really Said. *Proceedings of the Linnean Society of London* **168**: 70-87.

______ . 1959. *Lamarck and Modern Genetics.* Manchester, UK: University of Manchester Press.

Capra, Fritjof, and Pier Luigi Luisi. 2014. *The Systems View of Life: A Unifying Vision.* Cambridge, UK: Cambridge University Press.

Carnegie, Andrew. 1992/1889. Wealth, reprinted in *The Andrew Carnegie Reader,* ed. J.F. Wall, 129-154. Pittsburgh, PA: University of Pittsburgh Press.

Carneiro, Robert L. 1970. A Theory of the Origin of the State. *Science* **169**: 733-738.

______ . 1978. Political Expansion as an Expression of the Principle of Competitive Exclusion. In *Origins of the State,* ed. Ronald N. Cohen and Elman R. Service, 205-223. Philadelphia, PA: ISHI.

______ . 1981. The Chiefdom: Precursor of the State. In *The Transition to Statehood in the New World*, ed. Grant D. Jones, and Robert R. Kautz, 37-79. Cambridge, UK: Cambridge University Press.

Carrapiço, Francisco. 2010. How Symbiogenic is Evolution? *Theory in Bioscience* **129**: 135-139.

Carter, Charles W., Jr., and Richard Wolfenden. 2015. tRNA Acceptor Stem and Anticodon Bases Form Independent Codes Related to Protein Folding. *Proceedings of the National Academy of Sciences* **112**: 7489-7494.

Cavalier-Smith, Thomas. 1981. The Origin and Early Evolution of the Eukaryote Cell. *Symposia of the Society of General Microbiology* **36**: 33-84.

______ . 1991. The Evolution of Cells. In *Evolution of Life: Fossils, Molecules and Culture*, ed. Syozo Osawa, and Tasaku Honjo, 271-304. Tokyo: Springer Verlag.

______ . 2009. Predation and Eukaryote Cell Origins: A Coevolutionary Perspective. *The International Journal of Biochemistry & Cell Biology* **41**: 307-322.

______ . 2013. Symbiogenesis: Mechanisms, Evolutionary Consequences, and Systematic Implications. *Annual Review of Ecology and Evolutionary Systematics* **44**: 145-172.

Cavalli-Sforza, Luigi. Luca, and Marcus Feldman. 1981. *Cultural Transmission and Evolution: A Quantitative Approach*. Princeton, NJ: Princeton University Press.

Cavalli-Sforza, Luigi, Paolo Menozzi, and Alberto Piazza. 1994. *The History and Geography of Human Genes*. Princeton, NJ: Princeton University Press.

Chapais, Bernard. 2015. Competence and the Evolutionary Origins of Status and Power in Humans. *Human Nature* **26**: 161-183.

Cheney, Dorothy L., and Richard W. Wrangham. 1987. Predation. In *Primate Societies*, ed. Barbara B. Smuts, Dorothy L. Cheney, Robert M. Seyfarth, and Richard W. Wrangham, 227-239. Chicago, IL: University of Chicago Press.

Cheng, Joey T., Jessica L. Tracy, Tom Foulsham, Alan Kingstone, and Joseph Henrich. 2013. Two Ways to the Top: Evidence That Dominance and Prestige Are Distinct Yet Viable Avenues to Social Rank and Influence. *Journal of Personality and Social Psychology* **104**: 103-125.

Childe, V. Gordon. 1951/1936. *Man Makes Himself*. New York: New American Library.

Chomsky, Noam. 2004. Language and Mind: Current Thoughts on Ancient Problems. Part I & Part II. In *Variation and Universals in Biolinguistics*, ed. Lyle Jenkins, 379-405. Amsterdam: Elsevier.

Ciochon, Russell L., and John G. Fleagle. 1993. *The Human Evolution Source Book*. Englewood Cliffs, NJ: Prentice Hall.

Clark, J. Desmond. 1992. African and Asian Perspectives on the Origins of Modern Humans. *Philosophical Transactions of the Royal Society B* **337**: 201-215.

Clark, J. Desmond, and John W. K. Harris. 1985. Fire and its Roles in Early Hominid Lifeways. *The African Archaeological Review* **3**: 3-27.

Clutton-Brock, Timothy H., and Geoff A. Parker. 1995. Punishment in Animal Societies. *Nature* **373**: 209-216.

Cochran, Gregory, and Henry Harpending. 2009. *The 10,000 Year Explosion: How Civilization Accelerated Human Evolution*. New York: Basic Books.

Cohen, Mark Nathan. 1977. *The Food Crisis in Prehistory: Overpopulation and the Origins of Agriculture.* New Haven, CT: Yale University Press.

Collard, Mark, April Ruttle, Briggs Buchanan, and Michael J. O'Brien. 2013. Population Size and Cultural Evolution in Nonindustrial Food-Producing Societies. *PLoS ONE* **8**(9) e7628; doi: 10.1371/journal.pone.0072628

Collard, Mark, Krist Vaesen, Richard Cosgrove, and Wil Roebroeks. 2016. The Empirical Case Against the 'Demographic Turn' in Palaeolithic Archaeology. *Philosophical Transactions of the Royal Society B* 371: 20150242; http://dx.doi.org/10.1098/rstb.2015.0242

Conradt, Larissa, and Timothy J. Roper. 2003. Group Decision-Making in Animals. *Nature* **421**: 155-158.

Conroy Glenn C., *et al.* 1998. Endocranial Capacity in an Early Hominid Cranium from Sterkfontein South Africa. *Science* **280**: 1730-1731.

Coppens, Yves, and Brigitte Senut. 1991. *Origine(s) de la Bipédie les Hominidés.* Paris: Cahiers de Paléanthropologie, Centre National de la Recherche Scientifique.

Corbalis, Michael C. 2003. *From Hand to Mouth: The Origins of Language.* Princeton, NJ: Princeton University Press.

Corning, Peter A. 1982. Durkheim and Spencer. *The British Journal of Sociology* **33**: 359-382.

______ . 1983. *The Synergism Hypothesis: A Theory of Progressive Evolution.* New York: McGraw-Hill.

______ . 1996. The Co-operative Gene: On the Role of Synergy in Evolution. *Evolutionary Theory* **11**: 183-207.

______ . 2003. *Nature's Magic: Synergy in Evolution and the Fate of Humankind.* New York: Cambridge University Press.

______ . 2005. *Holistic Darwinism: Synergy, Cybernetics and the Bioeconomics of Evolution.* Chicago: University of Chicago Press.

______ . 2007a. Synergy Goes to War: A Bioeconomic Theory of Collective Violence. *Journal of Bioeconomics* **9**: 104-144.

______ . 2007b. Control Information Theory: The 'Missing Link' in the Science of Cybernetics. *Systems Research and Behavioral Science* **24**: 297-311.

______ . 2011. *The Fair Society: The Science of Human Nature and the Pursuit of Social Justice.* Chicago: University of Chicago Press.

______ . 2013. Rotating the Necker Cube: A Bioeconomic Approach to Cooperation and the Causal Role of Synergy in Evolution. *Journal of Bioeconomics* **15**: 171-193.

______ . 2014. Evolution 'On Purpose': How Behaviour has Shaped the Evolutionary Process. *Biological Journal of the Linnean Society* **112**: 242-60.

______ . 2017. The Evolution of Politics: A Biological Approach. In *Handbook of Biology and Politics*, ed. Steven A. Peterson and Albert Somit, pp. 55-84. Cheltenham, UK: Edward Elgar.

Corning, Peter A., and Steven J. Kline. 1998a. Thermodyanmics, Information and Life Revisited, part I: To Be or Entropy. *Systems Research and Behavioral Science* **15**: 273-295.

______ . 1998b. Thermodyanmics, Information and Life Revisited, part II: Thermoeconomics and Control Information. *Systems Research and Behavioral Science* **15**: 453-482.

Corning, Peter A., and Eörs Szathmáry. 2015. 'Synergistic Selection': A Darwinian Frame for the Evolution of Complexity. *Journal of Theoretical Biology* **371**: 45-58.

Corsi, Pietro. 1988. *The Age of Lamarck: Evolutionary Theories in France 1790-1830*. Berkeley, CA: University of California Press.

Couzin, Iain. 2007. Collective Minds. *Nature* **445**: 715.

Cowlishaw, Guy 1994. Vulnerability to Predation in Baboon Populations. *Behaviour* **131**: 293-304.

Crespi, Bernard J. 2001. The Evolution of Social Behavior in Microorganisms. *Trends in Ecology and Evolution* **16**: 178-183.

Crick, Francis H.C. 1970. Central Dogma of Molecular Biology. *Nature* **227**: 561-563.

______. 1981. *Life, Its Origin and Nature*. New York: Simon and Schuster.

Crow, James F. 1979. Genes that Violate Mendel's Rules. *Scientific American* **240**(2): 134-146.

Currie, Cameron R. 2001. A Community of Ants, Fungi, and Bacteria: A Multilateral Approach to Studying Symbiosis. *Annual Reviews of Microbiology* **55**: 357-80.

Currie, Cameron R., James A. Scott, Richard C. Summerbell, and David Malloch. 1999. Fungus-Growing Ants Use Anti-Biotic-Producing Bacteria to Control Garden Parasites. *Nature* **398**: 701-703.

Curtis, Helena and N. Sue Barnes. 1989. *Biology* (5th edn.). New York: Worth Publishers.

Darnell, James E, Harvey F. Lodish, and David Baltimore. 1990. *Molecular Cell Biology* (2nd. edn.). New York: Scientific American Books.

Dart, Raymond. 1953. The Predatory Transition from Ape to Man. *International Anthropological and Linguistic Review* **1**: 201-219.

______. 1959. *Adventures with the Missing Link*. New York: Harper and Brothers.

Darwin, Charles R. 1968/1859. *On the Origin of Species by Means of Natural Selection, or the Preservation of Favoured Races in the Struggle for Life*. Baltimore: Penguin.

______. 1882/1859. *On the Origin of Species by Means of Natural Selection, or the Preservation of Favoured Races in the Struggle for Life* (6th edn.). London: John R. Murray.

______. 1874/1871. *The Descent of Man, and Selection in Relation to Sex*. New York: A.L. Burt.

______. 1965/1873. *The Expression of Emotions in Man and Animals*. London: John Murray.

Darwin, Francis. 1887. *The Life and Letters of Charles Darwin*. London: John Murray.

Dawkins, Richard. 1989/1976. *The Selfish Gene*. New York: Oxford University Press.

______. 1982. *The Extended Phenotype*. Oxford: W.H. Freeman.

Deacon, Terrence W. 1997. *The Symbolic Species: The Co-Evolution of Language and the Brain*. New York: W.W. Norton.

Deamer, David W., ed. 1978. *Light Transcending Membranes: Structure, Function and Evolution*. New York: Academic Press.

Deamer, David, W. 2011. *First Life: Discovering the Connections Between Stars, Cells, and How Life Began*. Berkeley, CA: University of California Press.

Deamer, David W., and Joan Oro. 1980. Role of Lipids in Prebiotic Structures. *Biosystems* **12**: 167-175.

Dediu, Dan, *et al.* 2013. Cultural Evolution of Language. In *Cultural Evolution: Society, Technology, Language, and Religion,* ed. Peter J. Richerson, and Morton H. Christiansen, 303-332. Cambridge, MA: MIT Press.

de Duve, Christian. 1996. The Birth of Complex Cells. *Scientific American* **274**(4): 50-57.

De Heinzelin, Jean, *et al.* 1999. Environment and Behavior of 2.5 Million-Year-Old Bouri Hominids. *Science* **284**: 625-629.

de Menocal, Peter B. 2004. African Climate Change and Faunal Evolution during the Pliocene-Pleistocene. *Earth and Planetary Science Letters* **220**: 3-24.

______. 2011. Climate and Human Evolution. *Science* **331**: 540-542.

Derex, Maxine, Marie-Pauline Beugin, Bernard Godelle, and Michel Raymond. 2013. Experimental Evidence for the Influence of Group Size on Cultural Complexity. *Nature* **503**: 389-391.

d'Errico, Francesco, and Chris B. Stringer 2011. Evolution, Revolution, or Saltation Scenario for the Emergence of Modern Culture. *Philosophical Transactions of the Royal Society B* **366**: 1060-1069.

de Waal, Frans B.M. 1982. *Chimpanzee Politics: Power and Sex Among Apes.* New York: Harper & Row.

______. 1996. *Good Natured: The Origin of Right and Wrong in Humans and Other Animals.* Cambridge, MA: Harvard University Press.

______. 1997. *Bonobo: The Forgotten Ape.* Berkeley, CA: University of California Press.

Diamond, Jared M. 1992. *The Third Chimpanzee: The Evolution and Future of the Human Animal.* New York: Harper Collins.

______. 1995. The Evolution of Human Inventiveness. In *What is Life? The Next Fifty Years,* ed. Michael P. Murphy, and Luke A.J. O'Neill, 45-46. Cambridge, UK: Cambridge University Press.

______. 1997. *Guns, Germs and Steel: The Fates of Human Societies*. New York: W.W. Norton.

______. 2005. *Collapse: How Societies Choose to Fail or Succeed.* New York: Viking.

Domínguez-Rodrigo, Manuel and Rebeca Barba. 2006. New Estimates of Tooth Marks and Percussion Marks from FLK Zinj, Olduvai Gorge (Tanzania): The Carnivore-Hominid-Carnivore Hypothesis Falsified. *Journal of Human Evolution* **50**: 170-194.

______. 2007. Five More Arguments to Invalidate the Passive Scavenging Version of the Carnivore-Hominid-Carnivore Model: A Reply to Blumenschine, *et al.* (2007a). *Journal of Human Evolution* **53**: 427-433.

Donald, Merlin 1991. *Origins of the Modern Mind: The Stages in the Evolution of Culture and Cognition.* Cambridge, MA: Harvard University Press.

Dor, Daniel, and Eva Jablonka. 2000. From Cultural Selection to Genetic Selection: A Framework for the Evolution of Language. *Selection* **1**: 33-55.

Dugatkin, Lee. 1999. *Cheating Monkeys and Citizen Bees.* New York: Free Press.

Dunbar, Robin I.M. 1988. *Primate Social Systems.* London: Croom Helm.

______. 1996. Determinants of Group Size in Primates: A General Model. In *Evolution of Social Behaviour Patterns in Primates and Man,* ed. Walter G. Runciman, John Maynard Smith, and Robin I. M. Dunbar, 35-58. Oxford: Oxford University Press.

______. 1998. The Social Brain Hypothesis. *Evolutionary Anthropology* **6**: 178-190.

______ . 2001. Brain on Two Legs: Group Size and the Evolution of Intelligence. In *Tree of Origin: What Primate Behaviour Can Tell Us About Human Social Evolution,* ed. Frans B.M. de Waal, 173-191. Cambridge, MA: Harvard University Press.

______ . 2009. The Social Brain Hypothesis and its Implications for Social Evolution. *Annals of Human Biology* **36**: 562-572.

Dunbar, Robin I.M., and Susanne Shultz. 2007. Evolution in the Social Brain. *Science* **317**: 1344-1347.

Durkheim, Emile. 1997/1893. *The Division of Labor in Society.* New York: The Free Press.

Eberhart, Russell C., Yuhui Shi, and James Kennedy. 2001. *Swarm Intelligence.* San Francisco, CA: Morgan Freeman.

Ehrlich, Paul R. 2000. *Human Natures: Genes, Culture and the Human Prospect.* Washington, DC: Island Press.

Ehrlich, Paul R., and Anne H. Ehrlich. 2012. Can a Collapse of Global Civilization be Avoided? *Proceedings of the Royal Society B* **280**: 20122845; http://dx.doi.org/10.1098/rspb.2012.2845

Ehrlich, Paul R., and John Harte. 2015. Opinion: To Feed the World in 2050 Will Require a Global Revolution. *Proceedings of the National Academy of Sciences* **112**: 14743-14744.

Eigen, Manfred, and Peter Schuster. 1977. The Hypercycle: A Principle of Natural Self-Organization. *Die Naturwissenschaften* **64**(11): 541-65.

Eigen, Manfred, and PeterSchuster. 1979. *The Hypercycle: A Principle of Natural Self-Organization.* Berlin: Springer-Verlag.

El-Naggar, Mohamed Y., *et al.* 2010. Electrical Transport Along Bacterial Nanowires from *Shewanella oneidensis* MR-1. *Proceedings of the National Academy of Sciences* **107**: 18127-18131.

Ereshefsky, Marc, and Makmiller Pedroso. 2015. Rethinking Evolutionary Individuality. *Proceedings of the National Academy of Sciences* **112**: 1012-10132.

Fagan, Brian M. 1998. *People of the Earth: An Introduction to World Prehistory* (9th edn.). New York: Addison Wesley Longman, Inc.

Falk, Armin, Ernst Fehr, and Urs Fischbacher. 2001. Driving Forces of Informal Sanctions. Working Paper No. 59. Working Paper Series ISSN 1424-0429. Zurich, Switzerland: University of Zurich.

Farine, Damien R., Pierre-Olivier Montiglio, and Orr Spiegel. 2015. From Individuals to Groups and Back: The Evolutionary Implications of Group Phenotypic Composition. *Trends in Ecology & Evolution* **30**: 609-621.

Farmer, Malcolm. 1994. The Origins of Weapon Systems. *Current Anthropology* **35**: 679-681.

Fehr, Ernst., and Simon Gächter. 2000a. Cooperation and Punishment in a Public Goods Experiment. *American Economic Review* **90**: 980-994.

Fehr, Ernst, and Simon Gächter. 2000b. Fairness and Retaliation: The Economics of Reciprocity. *Journal of Economic Perspectives* **14**: 159-181.

Fehr, Ernst, Urs Fischbacher, and Simon Gächter. 2002. Strong Reciprocity, Human Cooperation, and the Enforcement of Social Norms. *Human Nature* **13**: 1-25.

Fernández, Eliseo. 2016. Synergy of Energy and Semiosis: Cooperation Climbs the Tree of Life. *Biosemiotics* **9**: 383-397.

Field, Katie J., Silvia Pressel, Jeffrey G. Duckett, William R. Rimington, and Martin I. Bidartondo. 2015. Symbiotic Options for the Conquest of Land. *Trends in Ecology & Evolution* **30**: 477-486.

Fisher, Simon E. and Matt Ridley. 2013. Culture, Genes, and the Human Revolution. *Science* **340**: 929-930.

Flannery, Kent, and Joyce Marcus. 2012. *The Creation of Inequality: How our Prehistoric Ancestors Set the Stage for Monarchy, Slavery, and Empire.* Cambridge, MA: Harvard University Press.

Fletcher, Jeffrey A., and Martin Zwick. 2004. Strong Altruism Can Evolve in Randomly Formed Groups. *Journal of Theoretical Biology* **228**(3): 303-313.

Foley, Robert A. 1994. Speciation, Extinction and Climatic Change in Hominid Evolution. *Journal of Human Evolution* **26**(4): 275-289.

______ . 1995. *Humans Before Humanity: An Evolutionary Perspective.* Oxford: Blackwell Publishers.

Fonseca-Azevedo, Karina, and Suzana Herculano-Houzel. 2012. Metabolic Constraint Imposes Tradeoff Between Body Size and Number of Brain Neurons in Human Evolution. *Proceedings of the National Academy of Sciences* **109**: 18571-18576.

______ . 1997. Models of Symbiosis. *The American Naturalist* **150**: S80-S99.

Franks, Nigel R. 1989. Army Ants: A Collective Intelligence. *American Scientist* **77**(2): 139-45.

Frederickson, Megan E. 2013. Rethinking Mutualism Stability: Cheaters and the Evolution of Sanctions. *Quarterly Review of Biology* **88**: 269-295.

Frohlich, Norman, and Joe A. Oppenheimer. 1992. *Choosing Justice: An Experimental Approach to Ethical Theory.* Berkeley, CA: University of California Press.

Gánti, Tibor. 2003/1971. *The Principles of Life.* Oxford: Oxford University Press.

______ . 2003. *Chemoton Theory: Theory of Living Systems.* Kluwer Academic/Plenum Publishers.

Gavrilets, Sergey, and Peter J. Richerson. 2017. Collective Action and the Evolution of Social Norm Internalization. Pnas.org/cgi/doi/10.1073/pnas.1703857114.

Ghiselin, Michael T. 1997. *Metaphysics and the Origin of Species.* Albany, NY: SUNY Press.

Gibson, Kathleen, and Tim Ingold, eds. 1993. *Tools, Language, and Cognition in Human Evolution.* Cambridge, UK: Cambridge University Press.

Gilroy, Simon and Anthony Trewavas. 2001. Signal Processing and Transduction in Plant Cells: The End and the Beginning. *Nature Reviews (Molecular Cell Biology)* **2**: 307-314.

Ginsburg, Simona, and Eva Jablonka. 2010. The Evolution of Associative Learning: A Factor in the Cambrian Explosion. *Journal of Theoretical Biology* **266**: 11-20.

Gintis, Herbert. 2000a. *Game Theory Evolving: A Problem-centered Introduction to Modeling Strategic Behavior.* Princeton, NJ: Princeton University Press.

______ . 2000b. Strong Reciprocity and Human Sociality. *Journal of Theoretical Biology* **206**: 169-179.

______ . 2007. The Evolution of Private Property. *Journal of Economic Behavior & Organization* **64**: 1-16.

______ . 2016. *Individuality and Entanglement: The Moral and Material Bases of Human Social Life.* Princeton, NJ: Princeton University Press.

Gintis, Herbert, Samuel Bowles, Robert Boyd, and Ernst Fehr. 2003. Explaining Altruistic Behavior in Humans. *Evolution and Human Behavior* **24**: 153-172.

Gintis, Herbert, Carel van Schaik, and Christopher Boehm. 2015. *Zoon Politikon*: The Evolutionary Origins of Human Political Systems. *Current Anthropology* **56**: 327-353.

Gissis, Snait B., and Eva Jablonka, eds. 2011. *Transformations of Lamarckism. From Subtle Fluids to Molecular Biology*. Cambridge, MA: MIT Press.

Glasser, Matthew F. *et al.* 2016. A Multi-Modal Parcellation of Human Cebebral Cortex. *Nature* **536**: 171-178.

Godfrey-Smith, Peter. 2015. Reproduction, Symbiosis, and the Eukaryotic Cell. *Proceedings of the National Academy of Sciences* **112**: 10120-10125.

Gold, Thomas. 1992. The Deep Hot Biosphere. *Proceedings of National Academy of Sciences* **89**(13): 6045-6049.

______. 1999. *The Deep Hot Biosphere*. New York: Springer..

Gorbach Sherwood L. 1990. Lactic Acid Bacteria and Human Health. *Annals of Medicine* **22**(1): 37-41.

Gordon, Deborah M. 2007. Control Without Hierarchy. *Nature* **446**: 143.

Goren-Inbar, Naama. 2011. Culture and Cognition in the Acheulian Industry: A Case Study from Gesher Yenot Yaʿqov. *Philosophical Transactions of the Royal Society B* **366**: 1038-1049.

Gould, James L., and Carol G. Gould. 1995. *The Honey Bee*. New York: Scientific American Library.

Gould, Stephen Jay. 1999a. A Division of Worms. *Natural History* **108**(1): 18.

______ . 1999b. Branching Through a Wormhole: Lamarck's Ladder Collapses. *Natural History* **108**(2): 24-27.

______ . 2002. *The Structure of Evolutionary Theory*. Cambridge, MA: Belknap (Harvard University Press).

Gowdy, John M., ed. 1998. *Limited Wants, Unlimited Means: A Reader on Hunter-Gatherer Economics and the Environment*. Washington, D.C.: Island Press.

Gowdy, John, and Lisi Krall. 2014. Agriculture as a Major Evolutionary Transition to Human Ultrasociality. *Journal of Bioeconomics* **16**(2): 179-202.

Gowdy, John, and Lisi Krall. 2015. The Economic Origins of Human Ultrasociality. *Behavioral and Brain Sciences* Jan; 39:e92. doi: 10.1017/S0140525X1500059X. Epub 2015 Apr 27.

Gowdy, John M., Denise E. Dollimore, David Sloan Wilson, and Ulrich Witt. 2013. Economic Cosmology and the Evolutionary Challenge. *Journal of Economic Behavior & Organization* **90S** S11-S20.

Gowlett, John A. J., and Richard W. Wrangham. 2013. Earliest Fire in Africa: Towards the Convergence of Archaeological Evidence and the Cooking Hypothesis. *Azania: Archaeological Research in Africa* **48**(1): 5-30.

Grant, Peter R., and B. Rosemary Grant. 2014. *40 Years of Evolution: Darwin's Finches on Daphne Major Island*. Princeton, NJ: Princeton University Press.

Guala, Francesco. 2009. Reciprocity: Weak or Strong? What Punishment Experiments Do (and Do Not) Demonstrate. *Behavioral and Brain Sciences* **35**: 1-59.

Haldane, John B. S. 1932. *The Causes of Evolution*. New York: Harper & Row.

Haldane, J.B.S., and Julian Huxley. 1927. *Animal Biology*. Oxford: Oxford University Press.

Halstead, Bruce W. 1988. *Poisonous and Venomous Marine Animals of the World*. London: Darwin Press.

Hamilton, William D. 1964a. The Genetical Evolution of Social Behavior, I. *Journal of Theoretical Biology* **7**: 1-16.

______ . 1964b. The Genetical Evolution of Social Behavior, II. *Journal of Theoretical Biology* **7**: 17-52.

Hardin, Garrett. 1968. The Tragedy of the Commons. *Science* **162**: 1243-1248.

Hardison, Ross. 1999. The Evolution of Hemoglobin. *American Scientist* **87**: 126-137.

Harmand, Sonia, *et al.* 2015. 3.3-million-year-old Stone Tools from Lomekwi 3, West Turkana, Kenya. *Nature* **521**: 310-315.

Hawkes, Kristin. 2003. Grandmothers and the Evolution of Human Longevity. *American Journal of Human Biology* **15**(3): 380-400.

Hawks, John, Eric T. Wang, Gregory M. Cochran, Henry C. Harpending, and Robert K. Moyzis. 2007. Recent Acceleration of Human Adaptive Evolution. *Proceedings of the National Academy of Sciences* **104**: 20753-20758.

Heckman, Daniel S., *et al.* 2001. Molecular Evidence for the Early Colonization of Land by Fungi and Plants. *Science* **293**: 1129-1133.

Heinrich, Berndt. 1995. An Experimental Investigation of Insight in Common Ravens (*Corvus corax*). *Auk:* **112**: 994-1003.

______ . 1999. *Mind of the Raven: Investigations and Adventures with Wolf-Birds.* New York: Harper Collins.

Hendry, Andrew P., Kiyoko M. Gotanda, and Erik I. Svensson. 2016. Human Influences on Evolution, and the Ecological and Societal Consequences. *Philosophical Transactions of the Royal Society B* 2017 372 20160028; doi: 10.1098/rstb.2016.0028.

Henrich, Joseph. 2016. *The Secret of Our Success: How Culture is Driving Human Evolution, Domesticating Our Species, and Making Us Smarter.* Princeton, N.J. and Oxford, UK: Princeton University Press.

Herculano-Houzel, Suzana. 2012. The Remarkable, Yet Not Extraordinary, Human Brain as a Scaled-up Primate Brain and its Associated Cost. *Proceedings of the National Academy of Sciences* **109**: 10661-10668.

Higgs, Paul G., and Niles Lehman. 2015. The RNA World; Molecular Cooperation at the Origins of Life. *Nature Reviews* (*Genetics*) **16**: 7-17.

Hill, Andrew H. 1987. Causes of Perceived Faunal Change in the Later Neogene of East Africa. *Journal of Human Evolution* **16**(7/8): 583-596.

Hill, Kim R. *et al.* 2011. Co-Residence Patterns in Hunter-Gatherer Societies Show Unique Human Social Structure. *Science* **331**: 1287-1289.

Hodgson, Geoffrey M. 2015. *Conceptualizing Capitalism: Institutions, Evolution, Future.* Chicago: University of Chicago Press.

Holland, John H. 1998. *Emergence: From Chaos to Order.* Reading, MA: Addison-Wesley (Helix Books).

Hölldobler, Bert, and Edward O. Wilson. 1990. *The Ants.* Cambridge, MA: Harvard University Press.

Hölldobler, Bert, and Edward O. Wilson. 1994. *Journey to the Ants,* Cambridge, MA: Belknap Press (Harvard University Press).

Hölldobler, Bert, and Edward O. Wilson. 2009. *The Superorganism: The Beauty, Elegance and Strangeness of Insect Societies.* New York: W.W. Norton.

Holloway, Ralph L. 1975. Early Hominid Endocasts: Volumes, Morphology, and Significance for Hominid Evolution. In *Primate Functional Morphology and Evolution,* ed. Russell Tuttle, 393-416. The Hague/ Paris: Mouton.

______ . 1983a. Cerebral Brain Endocast Pattern of *Australopithecus afarensis* Hominid. *Nature* **303**: 420-422.

______ . 1983b. Human Paleontological Evidence Relevant to Language Behavior. *Human Neurobiology* **2**: 105-114.

______ . 1996. Evolution of the Human Brain. In *Handbook of Human Symbolic Evolution,* ed. Andrew Lock, and Charles R. Peters, 74-116. Oxford: Oxford Science Publications.

______ . 1997. Brain Evolution. In *Encyclopedia of Human Biology,* (Vol. 2, 2nd edn.), ed. Renato Dulbecco, 189-200. New York: Academic Press.

Hordijk Wim, Stuart A. Kauffman, and Mike Steel. 2011. Required Levels of Catalysis for Emergence of Autocatalytic Sets in Models of Chemical Reaction Systems. *International Journal of Molecular Science* **12**(5): 3085-3101.

Hoyle, Fred. 1983. *The Intelligent Universe.* London: Michael Joseph Limited.

Hrdy, Sarah Blaffer. 2000. *Mother Nature: Maternal Instincts and How They Shape the Human Species.* New York: Ballantine.

______ . 2009. *Mothers and Others: The Evolutionary Origins of Mutual Understanding.* Cambridge, MA: Harvard University Press.

Hublin, Jean-Jacques, *et al.* 2017. New Fossils from Jebel Irhoud, Morocco and the Pan-African Origin of *Homo sapiens. Nature* **546**: 289-292.

Hunt, Paul. 1998. The Function of the Hox Genes. In *Developmental Biology,* ed. Edward E. Bittar, and Neville Bittar, 261-291. London: Elsevier.

Hutchinson, G. Evelyn. 1965. *The Ecological Theater and the Evolutionary Play.* New Haven, CT: Yale University Press.

Huxley, Julian S. 1942. *Evolution: The Modern Synthesis.* New York: Harper & Row.

Isaac, Glynn L. 1978. The Food Sharing Behavior of Proto Human Hominids. *Scientific American* **238**(4): 90-108.

______ . 1981. Emergence of Human Behaviour Patterns. *Philosophical Transactions of the Royal Society B* **292**: 177-188.

______ . 1983. Aspects of Human Evolution. In *Evolution from Molecules to Men,* ed. Derek Bendell, 509-544. New York: Cambridge University Press.

Isack, Hussein A., and Hans-Ulrich Reyer. 1989. Honeyguides and Honey Gatherers: Interspecific Communication in a Symbiotic Relationship. *Science* **243**: 1343-1346.

Isbell, Lynne A., and Truman P. Young. 1996. The Evolution of Bipedalism in Hominids and Reduced Group Size in Chimpanzees: Alternative Responses to Decreasing Resource Availability. *Journal of Human Evolution* **30**: 389-397.

Iwamoto, Toshitaka, Akio Mori, Masao Kawai, and Afework Bekele. 1996. Anti-Predator Behavior of Gelada Baboons. *Primates* **37**: 389-397.

Jablonka, Eva. 2002. Information: Its Interpretation, Its Inheritance, and Its Sharing. *Philosophy of Science* **69**: 578-605.

______ . 2004. From Replicators to Hereditably Varying Phenotypic Traits: The Extended Phenotype Revisited. *Biology and Philosophy* **19**: 353-375.

______ . 2013. Epigenetic Inheritance and Plasticity: The Responsive Germline. *Progress in Biophysics and Molecular Biology* **111**: 99-107.

Jablonka, Eva, and Marion J. Lamb. 1995. *Epigenetic Inheritance and Evolution: The Lamarckian Dimension.* New York: Oxford University Press.

Jablonka, Eva, and Marion J. Lamb. 2006. The Evolution of Information in the Major Transitions. *Journal of Theoretical Biology* **239**: 236-246.

Jablonka, Eva, and Gal Raz. 2009. Transgenerational Epigenetic Inheritance: Prevalence, Mechanisms, and Implications for the Study of Heredity and Evolution. *Quarterly Review of Biology* **84**(2): 131-176.

Jablonka, Eva, and Marion J. Lamb. 2014. *Evolution in Four Dimensions: Genetic, Epigenetic, Behavioral, and Symbolic Variation in the History of Life* (revised edition). Cambridge, MA: MIT Press.

Jablonka, Eva, Marion J. Lamb, and Eytan Avital. 1998. 'Lamarckian' Mechanisms in Darwinian Evolution. *Trends in Ecology and Evolution* **13**(5): 206-210.

Jaffe, Klaus. 2001. On the Relative Importance of Haplo-Diploidy, Assortative Mating and Social Synergy on the Evolutionary Emergence of Social Behavior. *Acta Biotheoretica* **49**: 29-42.

_______ . 2010. Quantifying Social Synergy in Insect and Human Societies. *Behavioral Ecology and Sociobiology* **64**: 1721-1724.

______ . 2016. Extended Inclusive Fitness Theory: Synergy and Assortment Drives the Evolutionary Dynamics in Biology and Economics. *SpringerPlus* **5**(1) 1-19, 2016, **5**: 1092; doi: 10.1186/s40064-016-2750-z

Jaffe, Klaus, and Gerardo Febres. 2016. Defining Synergy Thermodynamically using Quantitative Measurements of Entropy and Free Energy. *Complexity* **21**(S2): 235-242.

John, E. Roy, Phyllis Chesler, Frank Bartlett, and Ira Victor. 1968. Observation Learning in Cats. *Science,* **159**: 1489-1491.

Jurmain, Robert, Lynn Kilgore, and Wenda Trevathan. 2008. *Essentials of Physical Anthropology*. Boston: Cengage Learning.

Kaplan, Hillard S., Paul L. Hooper, and Michael Gurven. 2009. The Evolutionary and Ecological Roots of Human Social Organization. *Philosophical Transactions of the Royal Society B* **364**: 3289-3299.

Kauffman, Stuart A. 1971. Cellular Homeostasis, Epigenesis and Replication in Randomly Aggregated Macro-Molecular Systems. *Journal of Cybernetics* **1**: 71-96.

______ . 1986. Autocatalytic Sets of Proteins. *Journal of Theoretical Biology* **119**: 1-24.

______ . 1995. *At Home in the Universe: The Search for the Laws of Self-Organization and Complexity*. New York: Oxford University Press.

______ . 2000. *Investigations*. New York: Oxford University Press.

______ . 2008. *Reinventing the Sacred: A New View of Science, Reason, and the Sacred*. New York: Basic Books.

Keeling, Charles D., and Timothy P. Whorf. 2000. The 1800-year Oceanic Tidal Cycle: A Possible Cause of Rapid Climate Change. *Proceedings of the National Academy of Sciences* **97**: 1314-1319.

Kelly, Gavin, Dominic Kelly, and Andrew Gamble, eds. 1997. *Stakeholder Capitalism*. New York: St. Martin's Press.

Kerr, Richard A. 1996. New Mammalian Data Challenge Evolutionary Pulse Theory. *Science* **273**: 431-432.

Khakhina, Liya Nikolaevna. 1992a. Evolutionary Significance of Symbiosis: Development of the Symbiogenesis Concept. *Symbiosis* **14**: 217-228.

______ . 1992b. *Concepts of Symbiogenesis: A Historical and Critical Study of Russian Botanists*, (Lynn Margulis and Mark McMenamin, eds.) New Haven: Yale University Press.

Kiers, E. Toby, and Stuart A. West. 2015. Evolving New Organisms via Symbiosis: When and How Do Symbiotic Partnerships Become New, Integrated Organisms? *Science* **348**: 392-394.

Kingdon, Jonathan. 1993. *Self-Made Man: Human Evolution from Eden to Extinction?* New York: John Wiley & Sons.

______ . 2003. *Lowly Origin: Where, When, and Why Our Ancestors First Stood Up*. Princeton, NJ and Oxford, UK: Princeton University Press.

Kivell, Tracy L., and Daniel Schmitt. 2009. Independent Evolution of Knuckle-walking in African Apes Shows that Humans Did Not Evolve from a Knuckle-walking Ancestor. *Proceedings of the National Academy of Sciences* **106**: 4241- 4246.

Klein, Richard G. 1999. *The Human Career: Human Biological and Cultural Origins*, (2nd edn.). Chicago: University of Chicago Press.

Klein, Richard G. 2000. Archeology and the Evolution of Human Behavior. *Current Anthropology* **9**: 17-36.

______ . 2009. *The Human Career: Human Biological and Cultural Origins* (3rd edn.). Chicago: University of Chicago Press.

Koestler, Arthur. 1967. *The Ghost in the Machine*. New York: Macmillan.

Kohlberg, Lawrence. 1981. *Essays on Moral Development, Vol. I: The Philosophy of Moral Development*. San Francisco, CA: Harper & Row.

Kohlberg, Lawrence, Charles Levine, and Alexandra Hewer. 1983. *Moral Stages: A Current Formulation and a Response to Critics*. Basel, NY: Karger.

Koonin, Eugene V., and William Martin. 2005. On the Origin of Genomes and Cells Within Inorganic Compartments. *Trends in Genetics* **21**: 647-654.

Korten, David. 2015/1995. *When Corporations Rule the World* (3rd edn.). Oakland, CA: Berrett-Koehler Publishers.

Kramer, Karen L., and Erik Otárola-Castillo. 2015. When Mothers Need Others: The Impact of Hominin Life History Evolution on Cooperative Breeding. *Journal of Human Evolution* **84**: 16-24.

Kuhn, Steven L., Mary C. Stiner, David S. Reese, and Erksin Gulec. 2001. Ornaments of the Earliest Upper Paleolithic: New Insights from the Levant. *Proceedings of the National Academy of Sciences* **98**: 7641-7646.

Kummer, Hans. 1968. *Social Organization of Hamadryas Baboons: A Field Study*. Chicago: University of Chicago Press.

Kun Ádám, *et al.* 2015. The Dynamics of the RNA World: Insights and Challenges. *Annals of the New York Academy of Sciences* **1341**: 75-95.

Kurtén, Bjorn 1984. *Not from the Apes*. New York: Columbia University Press.

Lack, David L. 1961/1947. *Darwin's Finches*. New York: Harper & Row.

Lagi, Marco, Yavni Bar-Yam, Karla Z. Bertrand, and Yaneer Bar-Yam. 2015. Accurate Market Price Formation Model with Both Supply-demand and Trend-following for Global Food Prices Providing Policy Recommendations. *Proceedings of the National Academy of Sciences* **112**: E6119-E6128.

Lahr, Marta Mirazón, *et al.* 2016. Inter-group Violence Among Early Holocene Hunter-gatherers of West Turkana, Kenya. *Nature* **529**: 394-398.

Laland Kevin N., F. John Odling-Smee, and Marc W. Feldman. 1999. Evolutionary Consequences of Niche Construction and Their Implications for Ecology. *Proceedings of the National Academy of Sciences* **96**: 10242-10247.

Laland Kevin N., John Odling-Smee, and Sean Myles. 2010. How Culture Shaped the Human Genome: Bringing Genetics and the Human Sciences Together. *Nature Reviews, Genetics* **11**: 137-148.

Laland, Kevin N., Kim Sterelny, John Odling-Smee, William Hoppitt, and Tobias Uller. 2011. Cause and Effect in Biology Revisited: Is Mayr's Proximate-Ultimate Dichotomy Still Useful? *Science* **334**: 1512-1516.

Laland, Kevin N., John Odling-Smee, William Hoppitt, and Tobias Uller. 2013. More on How and Why: Cause and Effect in Biology Revisited. *Biology and Philosophy* **28**(5): 719-745.

Laland, Kevin N., *et al.* 2015. The Extended Evolutionary Synthesis: Its Structure, Assumptions and Predictions. *Proceedings of the Royal Society B*, http://dx.doi.org/10.1098/rspb.2015.1019

Lamarck, Jean-Baptiste. 1914/1809. *Zoological Philosophy: An Exposition With Regard to the Natural History of Animals* (trans. H. Elliot.). London: Macmillan & Co., Ltd.

Lane, Nick. 2009. *Life Ascending: The Ten Great Inventions of Evolution*. New York: W.W. Norton.

_______ . 2014. Bioenergetic Constraints on the Evolution of Complex Life. *Cold Spring Harbor Perspectives in Biology* **6**(5), p. a015982

Lane, Nick, and William Martin. 2010. The Energetics of Genome Complexity. *Nature* **467**: 929-934.

Leach, Helen M. 2003. Human Domestication Reconsidered. *Current Anthropology* **44**: 349-368.

Lehmann, Laurent, and Laurent Keller. 2006. The Evolution of Cooperation and Altruism – A General Framework and a Classification of Models. *Journal of Evolutionary Biology* **19**: 1365-1376.

Leigh, Egbert Giles, Jr. 1977. How Does Selection Reconcile Individual Advantage with the Good of the Group? *Proceedings of the National Academy of Sciences* **74**: 4542-4546.

______ . 1983. When Does the Good of the Group Override the Advantage of the Individual? *Proceedings of the National Academy of Sciences* **80**: 2985-989.

______ . 1991. Genes, Bees and Ecosystems: The Evolution of a Common Interest Among Individuals. *Trends in Ecology and Evolution* **6**: 257-262.

______ . 1995. The Major Transitions of Evolution (book review). *Evolution* **49**: 1302-1306.

______ . 2010a. The Evolution of Mutualism. *Journal of Evolutionary Biology* **23**: 2507-2528.

______ . 2010b. The Group Selection Controversy. *Journal of Evolutionary Biology* **23**: 6-19.

Le Maho, Yvonne. 1977. The Emperor Penguin: A Strategy to Live and Breed in the Cold. *American Scientist* **65**: 680-693.

Leonard, William R., and Marcia L. Robertson. 1994. Evolutionary Perspectives on Human Nutrition: The Influence of Brain and Body Size on Diet and Metabolism. *American Journal of Human Biology* **6**: 77-88.

Leonard, William R., and Marcia L. Robertson. 1997. Comparative Primate Energetics and Hominid Evolution. *American Journal of Physical Anthropology* **102**(2): 265-281.

Leonard, William R., J. Josh Snodgrass, and Marcia L. Robertson. 2007. Effects of Brain Evolution on Human Nutrition and Metabolism. *Annual Review of Nutrition* **27**: 311-327.

Levinson, Stephen C. and Dan Dediu. 2013. The Interplay of Genetic and Cultural Factors in Ongoing Language Evolution. In *Cultural Evolution: Society, Technology, Language, and Religion,* ed. Peter J. Richerson, and Morton H. Christiansen, 219-232. Cambridge, MA: MIT Press.

Lewin, Roger. 1993. *Human Evolution: An Illustrated Introduction.* (3r74.d ed.). Boston: Blackwell Scientific.

______ . 1997. *Bones of Contention.* Chicago: University of Chicago Press.

Lieberman, Daniel. E. 2013. *The Story of the Human Body: Evolution, Health, and Disease.* New York: Pantheon Books.

Lieberman, Philip. 1998. *Eve Spoke: Human Language and Human Evolution.* New York: W.W. Norton.

Lilley, David M. J., and John Sutherland. 2011. The Chemical Origins of Life and its Early Evolution: An Introduction. *Philosophical Transactions of the Royal Society B* **366**: 2853-2856.

Linton, Sally. 1971. Woman the Gatherer: Male Bias in Anthropology. In *Women in Cross-Cultural Perspective*, ed. Sue Ellen Jacob, 17-22. Champaign, IL: University of Illinois Press.

Liu, Hua, Franck Prugnolle, Andrea Manica, and Francois Balloux. 2006. A Geographically Explicit Genetic Model of Worldwide Human-Settlement History. *American Journal of Human Genetics* **79**(2): 230-237.

Lloyd Morgan, Conwy. 1923. *Emergent Evolution.* London: Williams and Norgate.

______ . 1926. *Life, Mind and Spirit.* London: Williams and Norgate.

______ . 1933. *The Emergence of Novelty.* New York: Henry Holt and Co.

Locke, John L. 1970/1690. *Two Treatises of Government* (P. Laslett, ed.). Cambridge, MA: Harvard University Press.

Lombard, Marlize, and Laurel Phillipson. 2010. Indications of Bow and Stone-Tipped Arrow Use 64,000 Years Ago in KwaZulu-Natal, South Africa. *Antiquity* **84**: 635-648.

Love, Alan C. 2010. Rethinking the Structure of Evolutionary Theory for an Extended Synthesis. In *Evolution – The Extended Synthesis,* ed. Massimo Pigliucci, and Gerd B. Müller, 443-481. Cambridge, MA: MIT Press.

Lovejoy, C. Owen. 1981. The Origin of Man. *Science* **211**: 341-350.

______ . 2009. Reexamining Human Origins in Light of *Ardipithecus ramidus. Science* **326**: 74e1-74e8.

Lumsden, Charles J., and Edward O. Wilson. 1981. *Genes, Mind, and Culture: The Coevolutionary Process.* Cambridge, MA: Harvard University Press.

Lynch, Don. 1992. *Titanic: An Illustrated History.* Toronto, Canada: Madison Press.

MacLarnon, Ann M., and Gwen P. Hewitt. 1999. The Evolution of Human Speech: The Role of Enhanced Breathing Control. *American Journal of Physical Anthropology* **109**: 341-363.

Maddox, John. 1990. Cooperating Molecules in Biology. *Nature* **345**: 979-980. Maisels, Charles Keith. 1999. *Early Civilizations of the Old World.* London and New York: Routledge.

Malthus, Thomas R. 2008/1798. *An Essay on the Principle of Population.* Oxford: Oxford World's Classics.

Manson, Joseph H., and Richard W. Wrangham. 1991. Intergroup Aggression in Chimpanzees and Humans. *Current Anthropology* **32**(4): 369-390.

Marean, Curtis W. 2015. The Most Invasive Species of All. *Scientific American* **313**(2): 32-39.

Margulis, Lynn. 1970. *Origin of Eukaryotic Cells.* New Haven, CT: Yale University Press.

______. 1981. *Symbiosis in Cell Evolution.* San Francisco, CA: W.H. Freeman.

______ . 1990. Words as Battle Cries -- Symbiogenesis and the New Field of Endocytobiology. *BioScience* **40**(9): 673-77.

______. 1993. *Symbiosis in Cell Evolution* (2nd edn.). New York: W.H. Freeman.

______. 1998. *Symbiotic Planet: A New Look at Evolution.* New York: Basic Books.

Margulis, Lynn, and René Fester, eds. 1991. *Symbiosis as a Source of Evolutionary Innovation: Speciation and Morphogenesis.* Cambridge, MA. MIT Press.

Margulis, Lynn, and Mark McMenamin. 1990. Marriage of Convenience: The Motility of the Modern Cell May Reflect an Ancient Symbiotic Union. *Sciences* **30**(5): 30-38.

Margulis, Lynn, and Mark McMenamin, eds. 1993. *Concepts of Symbiogenesis: A Historical and Critical Study of the Research of Russian Botanists.* New Haven, CT: Yale University Press.

Margulis, Lynn, and Michael F. Dolan. 1998. Did Centrioles and Kinetosomes Evolve from Bacterial Symbionts? *Symbiosis* **26**: 199-204.

Margulis, Lynn, and Dorion Sagan. 1995. *What is Life?* New York: Simon & Shuster.

Margulis, Lynn, and Dorion Sagan. 2002. *Acquiring Genomes: A Theory of the Origins of Species.* New York: Basic Books.

Margulis, Lynn, Michael F. Dolan, and Ricardo Guerrero. 2000. The Chimeric Eukaryote Origin of the Nucleus, from the Karyomastigont in Amitochondriate Protists. *Proceedings of the National Academy of Sciences* **97**(13): 6954-6959.

Marshall, Alfred. 1890. *Principles of Economics.* London: Macmillan.

Martin, Robert D. 1981. Relative Brainsize in Terrestrial Vertebrates. *Nature* **293**: 57-60.

Martin, Robert D., and Ann M. MacLarnon. 1985. Gestation Period, Neonatal Size, and Maternal Investment in Placental Mammals. *Nature* **313**: 220-223.

Martin, William F., and Miklós Müller. 1998. The Hydrogen Hypothesis for the First Eukaryote. *Nature* **391**: 37-41.

Martin, William F., and Marek Mentel. 2010. The Origin of Mitochondria. *Nature Education* **3**(9): 58.

Martin, William F., and Michael J. Russell. 2003. On the Origins of Cells: An Hypothesis for the Evolutionary Transitions from Abiotic Geochemistry to Chemoautotrophic Prokaryotes, and from Prokaryotes to Nucleated Cells. *Philosophical Transactions of the Royal Society B* **358**: 59-85.

Martin, William F., and Michael J. Russell. 2007. On the Origin of Biochemistry at an Alkaline Hydrothermal Vent. *Philosophical Transactions of the Royal Society B* **362**: 1887-1925.

Martin, William F., John Baross, Deborah Kelley, and Michael J. Russell. 2008. Hydrothermal Vents and the Origin of Life. *Nature Reviews in Microbiology* **6**: 805-814.

Masters, Roger D. 2008. Historical Change and Evolutionary Theory: From Hunter-Gatherer Bands to States and Empires. *Politics and the Life Sciences* **26**(2): 46-74.

Mathieson, Iain, *et al.* 2015. Genome-wide Patterns of Selection in 230 Ancient Eurasians. *Nature* **528**: 499-503.

Maturana, Humberto R., and Francisco J. Varela. 1980/1973. *Autopoiesis and Cognition: The Realization of Living.* Dordrecht, Netherlands: Reidel.

Maynard Smith, John. 1964. Group Selection and Kin Selection. *Nature* **201**: 1145-1147.

______ . 1982a. The Evolution of Social Behavior - A Classification of Models. In *Current Problems in Sociobiology,* ed. The King's College Sociobiology Group, 28-44. Cambridge, UK: Cambridge University Press.

______ . 1982b. *Evolution and the Theory of Games.* Cambridge, UK: Cambridge University Press.

______ . 1983. Models of Evolution. *Proceedings of the Royal Society of London B* **219**: 315-325.

______ . 1998. *Evolutionary Genetics* (2nd edn.). Oxford: Oxford University Press.

Maynard Smith, John, and Eörs Szathmáry. 1995. *The Major Transitions in Evolution.* Oxford: Freeman Press.

Maynard Smith, John, and Eörs Szathmáry. 1999. *The Origins of Life: From the Birth of Life to the Origin of Language.* Oxford: Oxford University Press.

Mayr, Ernst. 1960. The Emergence of Evolutionary Novelties. In *Evolution after Darwin,* ed. Sol Tax, Vol. I, 349-380. Chicago: University of Chicago Press.

______ . 1961. Cause and Effect in Biology. *Science* **134**: 1501-1506.

______ . 1974. Teleological and Teleonomic: A New Analysis. In *Boston Studies in the Philosophy of Science,* (Vol. XIV), ed. Robert S. Cohen, and Max W. Wartofsky, 91-117. Boston: Reidel.

______ . 2001. *What Evolution Is.* New York: Basic Books.

Mazur, Suzan. 2014. *The Origin of Life Circus: A How to Make Life Extravaganza.* New York: Caswell Books.

Mcbrearty, Sally, and Alison S. Brooks. 2000. The Revolution that Wasn't: A New Interpretation of the Origin of Human Behavior. *Journal of Human Evolution* **39**: 453-562.

McGrew, William C. 1992. *Chimpanzee Material Culture: Implications for Human Evolution.* Cambridge, UK: Cambridge University Press.

McHenry, Henry M. 1992. How Big were Early Hominids? *Evolutionary Anthropology* **1**: 15-20.

McManus, Jerry F., Delia W. Oppo, and James L. Cullen. 1999. A 0.5-million-year Record of Millennial-scale Climate Variability in the North Atlantic. *Science* **283**: 971-975.

McPherron, Shannon P., *et al.* 2010. Evidence for Stone-tool-assisted Consumption of Animal Tissues Before 3.39 Million Years Ago at Dikika, Ethiopia. *Nature* **466**: 857-860.

McShea, Daniel W. 2015. Bernd Rosslenbroich: On the Origin of Autonomy; A New Look at the Major Transitions (book review). *Biology and Philosophy* **30**(3): 439-446.

McShea, Daniel W., and Robert N. Brandon. 2010. *Biology's First Law: The Tendency for Diversity and Complexity to Increase in Evolutionary Systems.* Chicago: University of Chicago Press.

McShea,Daniel W., and Carl Simpson. 2011. The Miscellaneous Transitions in Evolution. In *The Major Transitions in Evolution Revisited,* ed. Brett Calcott, and Kim Sterelny, 19-34. Cambridge, MA: MIT Press.

Mellars, Paul. 2006. Why Did Modern Human Populations Disperse from Africa ca. 60,000 Years Ago? A New Model. *Proceedings of the National Academy of Sciences* **103**: 9381-9386.

Mellars, Paul, and Jennifer C. French. 2011. Tenfold Population Increase in Western Europe at the Neandertal-to-Modern Human Transition. *Science* **333**: 623-627.

Mereschkovsky, Konstantin C. 1909. The Theory of Two Plasms as the Foundation of Symbiogenesis, A New Doctrine About the Origins of Organisms (in Russian). *Proceedings of the Imperial Kazan University* [USSR], **12**: 1-102.

Mesoudi, Alex. 2011. *Cultural Evolution: How Darwinian Theory Can Explain Human Culture and Synthesize the Social Sciences.* Chicago: University of Chicago Press.

_________, *et al.* 2013. Is Non-Genetic Inheritance Just a Proximate Mechanism? A Corroboration of the Extended Evolutionary Synthesis. *Biological Theory* **7**(3): 189-195.

Michod, Richard E. 1999. *Darwinian Dynamics, Evolutionary Transitions in Fitness and Individuality.* Princeton, NJ: Princeton University Press.

______ . 2007. Evolution of Individuality During the Transition from Unicellular to Multicellular Life. *Proceedings of the National Academy of Sciences* **104**: 8613-8618.

______ . 2011. Evolutionary Transitions in Individuality: Multicellularity and Sex. In *The Major Transitions in Evolution Revisited,* ed. Brett Calcott, and Kim Sterelny, 169-197. Cambridge, MA: MIT Press.

Michod, Richard E., and Matthew D. Herron. 2006. Cooperation and Conflict During Evolutionary Transitions in Individuality. *Journal of Evolutionary Biology* **19**: 1406-1409.

Miller, Stanley L. 1953. Production of Amino Acids Under Possible Primitive Earth Conditions. *Science* **117**: 528-529.

Miller, Stanley L., and Harold C. Urey. 1959. Organic Compound Synthesis on the Primitive Earth. *Science* **130**: 245-251.

Mitchell, Amir, *et al.* 2009. Adaptive Prediction of Environmental Changes by Microorganisms. *Nature* **460**: 220-224.

Mithen, Steven, J. 2003. *After the Ice: A Global Human History, 20,000-5000 BC.* London: Weidenfeld & Nicolson.

______ . 2006. *The Singing Neanderthals: The Origins of Music, Language, Mind and Body*. Cambridge, MA: Harvard University Press.

Morgan, Thomas J. H. *et al.* 2015. Experimental Evidence for the Co-evolution of Hominin Tool-Making, Teaching, and Language. *Nature Communications*, doi: 10.1038/ncomms7029

Morowitz, Harold J. 1978. Proton Semiconductors and Energy Transduction in Biological Systems. *American Journal of Physiology* **235**: R99-R114.

______ . 1981. Phase Separation, Charge Separation and Biogenesis. *Biosystems* **14**: 41-47.

______ . 1992. *Beginnings of Cellular Life: Metabolism Recapitulates Biogenesis.* New Haven, CT: Yale University Press.

Morowitz, Harold J., Bettina Heinz, and David W. Deamer. 1987. The Chemical Logic of a Minimum Protocell. *Origins of Life* **18**: 281-287.

Morton, Dudley J. 1927. Human Origin, Correlation of Previous Studies on Private Fect and Posture with Other Morphological Evidence. *American Journal of Physical Anthropology* **10**: 173-203.

Newman, Rosemary W. 1970. Why Man is Such a Sweaty and Thirsty Animal: A Speculative Review. *Human Biology* **42**: 12-27.

Nitecki, Matthew H., and Doris Nitecki, eds. 1993. *Evolutionary Ethics.* Albany, NY: State University of New York Press.

Noble, Denis. 2006. *The Music of Life.* Oxford: Oxford University Press.

______ . 2010. Letter to the Editor, *Physiology News* **78**: 43.

______ . 2012. A Theory of Biological Relativity: No Privileged Level of Causation. *Interface Focus* **2**: 55-64.

______ . 2013. Physiology is Rocking the Foundations of Evolutionary Biology. *Experimental Physiology* **98**(8): 1235-1243.

Noble, Denis, Eva Jablonka, Michael J. Joyner, Gerd B. Müller, and Stig W. Omholt. 2014. Evolution Evolves: Physiology Returns to Centre Stage. *Journal of Physiology* **592**(11): 2237-2244; doi/10.1113/jphysiol.2014.273151/epdf

Nowak, Martin A. 2006. Five Rules for the Evolution of Cooperation. *Science* **314**: 1560-1563.

______ . 2011. *Super Cooperators: Altruism, Evolution and Why We Need Each Other to Succeed* (with R. Highfield). New York: Free Press.

Nowak, Martin A., Corina E. Tarnita, and Edward O. Wilson. 2010. The Evolution of Eusociality. *Nature* **446**: 1057-1062.

O'Brien, Michael J., and Kevin N. Laland. 2012. Genes, Culture, and Agriculture: An Example of Human Niche Construction. *Current Anthropology* **53**: 434-470.

O'Connell James F., Kristen Hawkes, Karen D. Lupo, and Nicholas G. Blurton-Jones. 2002. Male Strategies and Plio-Pleistocene Archaeology. *Journal of Human Evolution* **43**: 831-872.

Odling-Smee F. John, Kevin N. Laland, and Marc W. Feldman. 1996. Niche Construction. *American Naturalist* **147**: 641-648.

Odling-Smee F. John, Kevin N. Laland, and Marc W. Feldman. 2003. *Niche Construction: The Neglected Process in Evolution.* Princeton, NJ: Princeton University Press.

Odling-Smee, F. John, Douglas H. Erwin, Eric P. Palkovacs, Marc W. Feldman, and Kevin N. Laland. 2013. Niche Construction Theory: A Practical Guide for Ecologists. *Quarterly Review of Biology* **88**: 3-28.

O'Hara, Ann M., and Fergus Shanahan. 2006. The Gut Flora as a Forgotten Organ. *EMBO Reports* **7**(7): 688-693.

Okasha, Samir. 2005. Multilevel Selection and the Major Transitions in Evolution. *Philosophy of Science* **72**: 1013-1025.

Ostrom, Elinor. 2009. A Polycentric Approach for Coping with Climate Change. Policy Research Paper 5095, The World Bank, Washington. D.C.

Palameta, Boris, and Louis K. Lefebvre. 1985. The Social Transmission of a Food-Finding Technique in Pigeons: What is Learned? *Animal Behaviour* **33**: 892-896.

Parish, Amy Randall. 1996. Female Relationships in Bonobos (*Pan paniscus*). *Human Nature* **7**: 61-96.

Parker, Christopher, H., Earl R. Keefe, Nicole M. Herzog, James F. O'Connell, and Kristen Hawkes. 2016. The Pyrophilic Primate Hypothesis. *Evolutionary Anthropology* **25**(2): 54-63.

Parsons. Talcott. 1949/1937. *The Structure of Social Action.* Chicago: The Free Press.

Patel, Bhavesh H., Claudia Percivalle, Dougal J. Ritson, Colm D. Duffy, and John D. Sutherland. 2015. Common Origins of RNA, Protein and Lipid Precursors in a Cyanosulfidic Protometabolism. *Nature Chemistry* **7**: 301-307.

Peace, William J. H., and Peter J. Grubb. 1982. Interaction of Light and Mineral Nutrient Supply in the Growth of *Impatiens Parviflora. The New Phytologist* **90**(1): 127-150.

Pendick, Daniel. 1997. Living Proof. *Earth* **6**(2): 24-25.

Pennisi, Elizabeth. 1999. Human Evolution: Did Cooked Tubers Spur the Evolution of Big Brains? *Science* **283**: 2004-2005.

Pereira, Luisa, Telma Rodrigues, and Francisco Carrapiço. 2012. A Symbiogenic Way in the Origin of Life. In *Genesis - In the Beginning: Precursors of Life, Chemical Models and Early Biological Evolution.* ed. Joseph Seckbach, 723-742. Dordrecht, Netherlands: Springer Science+Business Media.

Pfeiffer, John E. 1977. *The Emergence of Society: A Pre-history of the Establishment.* New York: McGraw-Hill.

Piaget, Jean. 1932. *The Moral Judgment of the Child.* London: Kegan Paul, Trench, Trubner Co.

Pickering, Travis. 2013. *Rough and Tumble: Aggression, Hunting, and Human Evolution.* Berkeley, CA: University of California Press.

Pigliucci Massimo, and Gerd B. Müller. 2010. *Evolution – The Extended Synthesis.* Cambridge, MA: MIT Press.

Piketty, Thomas. 2014. *Capital in the Twenty-First Century.* Cambridge, MA: The Belknap Press of Harvard University Press.

Pinker, Steven. 1994. *The Language Instinct.* New York: William Morrow and Co.

______ . 2010. The Cognitive Niche: Coevolution of Intelligence, Sociality, and Language. *Proceedings of the National Academy of Sciences* **107** (Suppl. 2): 8993-8999.

Pinker, Steven, and Paul Bloom. 1990. Natural Language and Natural Selection. *Behavioral and Brain Sciences* **13**(4): 707-784.

Plato. 1946/380B.C. *The Republic* (trans. B. Jowett). Cleveland, OH: World Publishing Company.

Plotkin, Henry C., ed. 1988. *The Role of Behavior in Evolution.* Cambridge, MA: MIT Press.

_____ . 2010. *Evolutionary Worlds Without End.* New York: Oxford University Press.

Pontzer, Herman, *et al.* 2010. Locomotor Anatomy and Biomechanics of the Dmanisi Hominins. *Journal of Human Evolution* **58**: 492-504.

Potts, Richard. 1984. Home Bases and Early Hominids. *American Scientist* **72**(4): 338-347.

______. 1988. *Early Hominid Activities at Olduvai.* Chicago: Aldine.

______ . 1996. *Humanity's Descent: The Consequences of Environmental Instability.* New York: Aldine DeGruyter.

______ . 1998a. Environmental Hypothesis of Hominid Evolution. *Yearbook of Physical Anthropology* **41**: 93-136.

______ . 1998b. Variability Selection in Hominid Evolution. *Evolutionary Anthropology* **7**(3): 81-96.

______ . 2012. Evolution and Environmental Change in Early Human Prehistory. *Annual Review of Anthropology* **41**: 151-167.

Potts, Richard, and Pat Shipman. 1981. Cutmarks Made by Stone Tools on Bones from Olduvai Gorge, Tanzania. *Nature* **291**: 557-580.

Potts, Richard, and J. Tyler Faith. 2015. Alternating High and Low Climate Variability: The Context of Natural Selection and Speciation in Plio-Pleistocene Hominin Evolution. *Journal of Human Evolution* **87**: 5-20.

Powell, Adam, Stephen Shennan, and Mark G. Thomas. 2009. Late Pleistocene Demography and the Appearance of Modern Human Behavior. *Science* **324**: 1298-1301.

Powner, Matthew W., Béatrice Gerland, and John D. Sutherland. 2009. Synthesis of Activated Pyrimidine Ribonucleotides in Prebiotically Plausible Conditions. *Nature* **459**: 239-242.

Price, Peter W. 1991. The Web of Life: Development Over 3.8 Billion years of Trophic Relationship. In *Symbiosis as a Source of Evolutionary Innovation,* ed., Lynn Margulis, and Rene Fester, 262-272. Cambridge, MA: MIT Press.

Puurtinen, Mikael, and Tapio Mappes. 2009. Between-Group Competition and Human Cooperation. *Proceedings of the Royal Society B* **276**: 355-360.

Queller, David C. 1997. Cooperators Since Life Began. (Review of *The Major Transitions in Evolution,* by John Maynard Smith and Eörs Szathmáry.) *Quarterly Review of Biology* **72**: 184-188.

______ . 2000. Relatedness and the Fraternal Major Transitions. *Philosophical Transactions of the Royal Society B* **355**: 1647-1655.

______ . 2004. Kinship is Relative. *Nature* **430**: 975-976.

Ratcliff, William C., R. Ford Denison, Mark Borrello, and Michael Travisano. 2012. Experimental Evolution of Multicellularity. *Proceedings of the National Academy of Sciences* **109**: 1595-1600.

Ratnieks, Francis L.W., and P. Kirk Visscher. 1989. Worker Policing in the Honey Bee. *Nature* **342**: 796-797.

Raven, John A. 1992. Energy and Nutrient Acquisition by Autotrophic Symbioses and their Asymbiotic Ancestors. *Symbiosis* **14**: 33-60.

Reynolds, Vernon, Vincent Falger, and Ian Vine. 1987. *The Sociobiology of Ethnocentrism: Evolutionary Dimensions of Xenophobia, Discrimination, Racism and Nationalism.* London: Croom Helm.

Richerson, Peter J. 2013. Group Size Determines Cultural Complexity. *Nature* **503**: 351-352.

Richerson, Peter J., and Robert Boyd. 1992. Cultural Inheritance and Evolutionary Ecology. In *Evolutionary Ecology and Human Behavior,* ed. Eric Alden Smith, and Bruce Winterhalder, 61-94. New York: Aldine de Gruyter.

Richerson, Peter J., and Robert Boyd. 1999. Complex Societies: The Evolutionary Origins of a Crude Superorganism. *Human Nature* **10**: 253-289.

Richerson, Peter J., and Robert Boyd. 2005. *Not by Genes Alone: How Culture Transformed Human Evolution*. Chicago: University of Chicago Press.

Richerson, Peter J., and Robert Boyd. 2010. Why Possibly Language Evolved. *Biolinguistics* **4**: 289-306.

Richerson, Peter J. and Robert Boyd. 2013. Rethinking Paleoanthropology: A World Queerer Than We Supposed. In Evolution of Mind, Brain and Culture, ed. Gary Hatfield, and Holly Pittman, 263-302. Penn Museum International Research Conferences Series, University of Pennsylvania Press.

Richerson, Peter J., and Morton H. Christiansen, eds. 2013. *Cultural Evolution: Society, Technology, Language, and Religion.* Cambridge, MA: MIT Press.

Richerson, Peter J., Robert Boyd, and Robert L. Bettinger. 2001. Was Agriculture Impossible During the Pleistocene but Mandatory During the Holocene? A Climate Change Hypothesis. *American Antiquity* **66**: 387-411.

Richerson, Peter J., Robert L. Bettinger, and Robert Boyd. 2005. Evolution on a Restless Planet: Were Environmental Variability and Environmental Change Major Drivers of Human Evolution? In *Handbook of Evolution: Evolution of Living Systems (Including Hominids),* ed. Franz M. Wuketits and Francisco J. Ayala, 223-242. Weinheim, Germany: Wiley-VCH.

Richerson, Peter J., Robert L. Bettinger, and Robert Boyd. 2009. Cultural Innovations and Demographic Change. *Human Biology* **81**: 211-235.

Richerson, Peter J., *et al.* 2016. Cultural Groups Selection Plays an Essential Role in Explaining Human Cooperation. *Behavioral and Brain Sciences* Vol. 39 e30.

Richmond, Brian G., David R. Begun, and David S. Strait. 2001. Origin of Human Bipedalism: The Knuckle-walking Hypothesis Revisited. *American Journal of Physical Anthropology.* Suppl **33**: 70-105.

Ricketts, Edward, and Jack Calvin. 1985. *Between Pacific Tides.* Stanford, CA: Stanford University Press.

Ridley, Mark. 2001. *The Cooperative Gene: How Mendel's Demon Explains the Evolution of Complex Beings.* New York: The Free Press.

Ridley, Matt. 2010. *The Rational Optimist: How Prosperity Evolves.* New York: HarperCollins.

Rilling, James K., and Thomas R Insel. 1999. The Primate Neocortex in Comparative Perspective Using Magnetic Resonance Imaging. *Journal of Human Evolution* **37**: 191-223.

Roberts, Alice. 2009. *The Incredible Human Journey: The Story of How We Colonised the Planet.* London: Bloomsbury Publishing.

Rodman, Peter S., and Henry M. McHenry. 1980. Bioenergetics and the Origins of Bipedalism. *American Journal of Physical Anthropology* **52**: 103-106.

Rose, Lisa, and Fiona Marshall. 1996. Meat Eating, Hominid Sociality, and Home Bases Revisited. *Current Anthropology* **37**(2): 307-338.

Rosen, Robert. 1991. *Life Itself: A Comprehensive Inquiry into the Nature, Origin, and Fabrication of Life.* New York: Columbia University Press.

Rowlett, Ralph. 1999. Did the Use of Fire for Cooking Lead to a Diet Change that Resulted in the Expansion of Brain Size in *Homo erectus* from that of *Australopithecus africanus*? *Science* **284**: 741.

Russell, Mike. 2006. First Life. *American Scientist* **94**: 32-39.

Russell, Nerissa. 2012. *Social Zooarchaeology: Humans and Animals in Prehistory.* Cambridge, UK: Cambridge University Press.

Sachs, Joel L., Ulrich G. Mueller, Thomas P. Wilcox, and James J. Bull. 2004. The Evolution of Cooperation. *Quarterly Review of Biology* **79**: 135-160.

Sahle, Yonatan, *et al.* 2013. Earliest Stone-tipped Projectiles from the Ethiopian Rift Date to More Than 279,000 Years Ago. *PLoS ONE* **8**(11): 1-9.

Salisbury, Richard F. 1962. *From Stone to Steel: Economic Consequences of a Technological Change in New Guinea.* Victoria, Australia: Melbourne University Press.

______. 1973. Economic Anthropology. *Annual Review of Anthropology* **2**: 85-94.

Schick, Kathy D., and Nicholas Toth. 1993. *Making Silent Stones Speak.* New York: Simon & Schuster.

Schleussner, Carl-Friedrich, Jonathan F. Donges, Reik V. Donner, and Hans Joachim Schellnhuber. Armed-Conflict Risks Enhanced by Climate-Related Disasters in Ethnically Fractionalized Countries. *Proceedings of the National Academy of Sciences* **113**: 9216-9221.

Schopf, William, *et al.* 2015. Sulfur-cycling Fossil Bacteria from the 1.8-Ga Duck Creek Formation Provide Promising Evidence of Evolution's Null Hypothesis. *Proceedings of the National Academy of Sciences* **112**: 2087-2092.

Schrödinger, Erwin. 1944. *What is Life? The Physical Aspect of the Living Cell.* Cambridge, UK: Cambridge University Press.

Schuppli,Caroline, Sereina M. Graber, Karin Isler, and Carel P. van Schaik. 2016. Life History, Cognition and the Evolution of Complex Foraging Niches. *Journal of Human Evolution* **92**: 91-100.

Seabright, Paul. 2004. *The Company of Strangers: A Natural History of Economic Life.* Princeton, NJ: Princeton University Press.

Sears Cynthia L. 2005. A Dynamic Partnership: Celebrating our Gut Flora. *Anaerobe* **11**(5): 247-251.

Seeley, Thomas D. 1989. The Honey Bee Colony as a Super-Organism. *American Scientist* **77**: 546-553.

______. 1995. *The Wisdom of the Hive: The Social Physiology of Honey Bee Colonies.* Cambridge, MA: Harvard University Press.

______. 2010. *Honeybee Democracy.* Princeton, NJ: Princeton University Press.

Semino, Ornella, *et al.* 2000. The Genetic Legacy of Paleolithic *Homo sapiens sapiens* in Extant Europeans: A Y Chromosome Perspective. *Science* **290**: 1155-1159.

Service, Elman. 1971. *Cultural Evolutionism: Theory in Practice.* New York; Holt, Rinehart and Winston.

Sethi, Rajiv, and Rohini Somanathan. 2001. Preference Evolution and Reciprocity. *Journal of Economic Theory* **97**: 273-297.

Sewall, Kendra B. 2015. Social Complexity as a Driver of Communication and Cognition. *Integrative and Comparative Biology* **3**: 384-395.

Shapiro, James A. 1988. Bacteria as Multicellular Organisms. *Scientific American* **258**(6): 82-89.

______ . 2009. Revisiting the Central Dogma in the 21st Century. *Annals of the New York Academy of Sciences* **1178**: 6-28.

______ . 2011. *Evolution: A View from the 21st Century.* Upper Saddle River, NJ: FT Press Science.

Shapiro, James A., and Martin Dworkin, eds. 1997. *Bacteria as Multicellular Organisms.* New York: Oxford University Press.

Shea, John J., and Matthew L. Sisk. 2010. Complex Projectile Technology and *Homo sapiens* Dispersal into Western Eurasia. *Paleoanthropology* **2010:** 100-122; doi: 10:4207/PA.2010.ART36

Sherman, Paul W., Jennifer U.M. Jarvis, and Richard D. Alexander, eds. 1991. *The Biology of the Naked Mole-Rat.* Princeton, NJ: Princeton University Press.

Sherman, Paul W., Jennifer U.M. Jarvis, and Stanton H. Braude. 1992. Naked Mole Rats. *Scientific American* **267**(2): 72-78.

Shipman, Pat. 1983. Early Hominid Lifestyle: Hunting and Gathering or Foraging and Scavenging. In *Animals and Archaeology*, ed. Juliet Clutton-Brook, and Caroline Grigson, 31-49. Oxford: British Archaeological Reports.

______ . 1986. Studies of Hominid-Faunal Interactions at Olduvai Gorge. *Journal of Human Evolution* **15**(8): 691-706.

Shipman, Pat, and Alan Walker. 1989. The Costs of Becoming a Predator. *Journal of Human Evolution* **18**: 373-392.

Shriver, Lionel. 2016. *The Mandibles: A Family 2029-2047.* New York: Harper/HarperCollins Publishers.

Shultz, Susanne, Christopher Opie, and Quentin D. Atkinson. 2011. Stepwise Evolution of Stable Sociality in Primates. *Nature* **479**: 219-222.

Shumaker, Robert W., Kristina R. Walkup, and Benjamin B. Beck. 2011. *Animal Tool Behavior: The Use and Manufacture of Tools by Animals.* Baltimore, MD: Johns Hopkins University Press.

Silk, Joan B. 2011. The Path to Sociality. *Nature* **49**: 182-183.

Simonti, Corinne, *et al.* 2016. The Phenotypic Legacy of Admixture Between Modern Humans and Neandertals. *Science* **351**: 737-741.

Simpson, Carl. 2012. The Evolutionary History of the Division of Labour. *Proceedings of the Royal Society B* **279**: 116-121.

Simpson, George Gaylord. 1953. The Baldwin Effect. *Evolution* **2**: 110-117.

Singer, Emily. 2014. Does Competition Drive Species Diversity? *Quanta Magazine* March 10, 2014.

Skyrms, Brian. 2004. *The Stag Hunt and the Evolution of Social Structure*. Cambridge, UK: Cambridge University Press.

Smaldino, Paul E. 2014. The Cultural Evolution of Emergent Group-level Traits. *Behavioral and Brain Sciences* **37**: 243-295.

Smith, Adam. 1976/1759. *The Theory of Moral Sentiments.* Oxford: Clarendon Press.

______ . 1964/1776. *The Wealth of Nations.* (2 Vols.) London: Dent.

Smith, Alex R., Rachel N. Carmody, Rachel J. Dutton, and Richard W. Wrangham. 2015. The Significance of Cooking for Early Hominin Scavenging. *Journal of Human Evolution* **84**: 62-70.

Snowdon, Charles T. 2001. From Primate Communication to Human Language. In *Tree of Origin: What Primate Behavior Can Tell Us About Human Evolution,* ed. Frans B.M. de Waal, 193-228. Cambridge, MA: Harvard University Press.

Sober, Elliott, and David Sloan Wilson. 1998. *Unto Others: The Evolution and Psychology of Unselfish Behavior.* Cambridge, MA: Harvard University Press.

Sockol, Michael D., David A. Raichlen, and Herman Pontzer. 2007. Chimpanzee Locomotor Energetics and the Origin of Human Bipedalism. *Proceedings of the National Academy of Sciences* **104**: 12265-12269.

Spalding Douglas A. 1873. Instinct: With Original Observations on Young Animals. *Macmillan's Magazine* **27**: 282-293.

Spencer, Herbert. 1852. A Theory of Population Deduced from the General Law of Animal Fertility. *Westminster Review* **LXII**: 468-501.

______. 1892/1852. The Development Hypothesis. In *Essays: Scientific, Political and Speculative.* New York: Appleton.

Spencer, Herbert. 1897. *The Principles of Sociology* (3rd edn.. 3 Vols.). New York: Appleton.

Spottiswoode, Claire N., Keith S. Begg, and Colleen M Begg. 2016. Reciprocal Signaling in Honeyguide-Human Mutualism. *Science* **353**: 387-389.

Spribille, Toby, *et al.* 2016. Basidiomycete Yeasts in the Cortex of Ascomycete Macrolichens. *Science* **353**: 488-492.

Stahlschmidt, Mareike C., *et al.* 2015. On the Evidence for Human Use and Control of Fire at Schöningen. *Journal of Human Evolution* **89**: 181-201.

Stanford, Craig B. 1999. *The Hunting Apes: Meat Eating and the Origins of Human Behavior.* Princeton, NJ: Princeton University Press.

Stanley, Steven M. 1992. Ecological Theory for the Origin of *Homo. Paleobiology* **18**: 237-257.

______ . 2000. The Past Climate Change Heats Up. *Proceedings of the National Academy of Sciences* **97**: 1319.

Steele, James. 1999. Stone Legacy of Skilled Hands. *Nature* **399**: 24-25.

Sterelny, Kim. 2012. Language, Gesture, Skill: The Co-Evolutionary Foundations of Language. *Philosophical Transactions of the Royal Society B* **367**: 2141-2151.

______ . 2013. Life in Interesting Times: Cooperation and Collective Action in the Holocene. In *Cooperation and Its Evolution*, ed. Kim Sterelny, Richard Joyce, Brett Calcott, and Ben Fraser, 89-108. Cambridge, MA: MIT Press.

Sterelny, Kim, Richard Joyce, Brett Calcott, and Ben Fraser, eds. 2013. *Cooperation and Its Evolution.* Cambridge, MA: MIT Press.

Steudel, Karen L. 1994. Locomotor Energetics and Hominid Evolution. *Evolutionary Anthropology* **3**: 42-48.

______ . 1996. Limb Morphology, Bipedal Gait, and the Energetics of Hominid Locomotion. *American Journal of Physical Anthropology* **99**: 345-355.

Stigler, George J. 1951. The Division of Labor is Limited by the Extent of the Market. *The Journal of Political Economy* **59**(3): 185-193.

Stock, Gregory. 2002. *Redesigning Humans: Our Inevitable Genetic Future*. New York: Houghton Mifflin.

Stoelhorst, J. Jan-Willem, and Peter J. Richerson. 2013. A Naturalistic Theory of Economic Organization. *Journal of Economic Behavior & Organization* **90S**: S45-S56.

Stringer, Chris. 2003. Human Evolution: Out of Ethiopia. *Nature* **423**: 692-695.

______. 2012. *The Origin of Our Species*. London: Penguin Books.

Strum, Shirley C. 1987. *Almost Human: A Journey into the World of Baboons*. New York: Random House.

Sumpter, David. 2010. *Collective Animal Behavior*. Princeton, NJ: Princeton University Press.

Suzuki, Toshikata N., David Wheatcroft, and Michael Griesser. 2016. Experimental Evidence for Compositional Syntax in Bird Calls. *Nature Communications* **7**, 10986; doi: 10.1038/ncomms10986 (8 March 2016).

Szathmáry, Eörs. 2002. Cultural Processes: The Latest Major Transition in Evolution. In *Encyclopedia of Cognitive Science,* ed. Lynn Nadel, London: Nature Publishing Group, Macmillan. doi: 10.1002/0470018860.s00716.

______. 2015. Toward Major Evolutionary Transitions Theory 2.0. *Proceedings of the National Academy of Sciences* **112**(33): 10104-10111.

Szathmáry, Eörs, and Szabolcs Számadó, 2008. Language: A Social History of Words. *Nature* **456**(6): 40-41.

Tagkopoulos, Ilias, Yir-Chung Liu, and Saeed Tavazoie. 2008. Predictive Behavior within Microbial Genetic Networks. *Science* **320**: 1313-1317.

Tainter, Joseph A. 1988. *The Collapse of Complex Societies.* New York: Cambridge University Press.

Tattersall, Ian. 1998. *Becoming Human: Evolution and Human Uniqueness.* New York: Harcourt Brace.

______. 2000. Once We Were Not Alone. *Scientific American* **282**(1): 56-62.

Taylor, Kendrick. 1999. Rapid Climate Change. *The American Scientist* **87**: 320-327.

Teaford, Mark F., and Peter S. Ungar. 2000. Diet and the Evolution of the Earliest Human Ancestors. *Proceedings of the National Academy of Sciences* **97**: 13506-13511.

Thieme, Hartmut. 1997. Lower Palaeolithic Hunting Spears from Germany. *Nature* **385**: 807-810.

Thomas, Chris D. 2015. Rapid Acceleration of Plant Speciation During the Anthropocene. *Trends in Ecology & Evolution* **30**: 448.455.

Thompson, Bill, Simon Kirby, and Kenny Smith. 2016. Culture Shapes the Evolution of Cognition. *Proceedings of the National Academy of Sciences* **113**: 4530-4535.

Thorne, Alan G., and Darren Curnoe. 2006. What is the Real Age of Adam and Eve? Proceedings of the Australian Society of Human Biology. *Homo-Journal of Comparative Human Biology* **57**: 240.

Thorpe, Susannah K.., Roger L. Holder, and Robin H. Crompton. 2007. Origin of Human Bipedalism as an Adaptation for Locomotion on Flexible Branches. *Science* **316**: 1328-1331.

Tobias, Philip V. 1971. *The Brain in Hominid Evolution.* New York: Columbia University Press.

_______ . ed. 1985. *Hominid Evolution: Past, Present and Future. Proceedings of the Taung Diamond Jubilee International Symposium.* New York: Alan R. Liss, Inc.

Tokuyama, Nahoko, and Takeshi Furuichi. 2016. Do Friends Help Each Other? Patterns of Female Coalition Formation in Wild Bonobos at Wamba. *Animal Behaviour* **119**: 27-35.

Tomasello, Michael. 2001. *The Cultural Origins of Human Cognition.* Cambridge, MA: Harvard University Press.

______ . 2005. *Constructing a Language: A Usage-Based Theory of Language Acquisition.* Cambridge, MA: Harvard University Press.

______ . 2008. *Origins of Human Communication.* Cambridge, MA: MIT Press.

______ . 2009. *Why We Cooperate.* Cambridge, MA: MIT Press (A Boston Review Book).

______ . 2014. *A Natural History of Human Thinking.* Cambridge, MA: Harvard University Press.

______ . 2016. *A Natural History of Human Morality.* Cambridge, MA: Harvard University Press.

Tomasello, Michael, Alicia P. Melis, Claudio Tennie, Emily Wyman, and Esther Herrmann. 2012. Two Key Steps in the Evolution of Human Cooperation: The Interdependence Hypothesis. *Current Anthropology* **53**(6): 673-692.

Toth, Nicholas. 1987a. Behavioral Influences from Early Stone Artifact Assemblages: An Experimental Model. *Journal of Human Evolution* **16**(7/8): 763-787.

______ . 1987b. The First Technology. *Scientific American* **255**(4): 112-121.

Toth, Nicholas, Kathy D. Schick, E. Sue Savage-Rumbaugh, Rose A. Sevcik, and Duane M. Rumbaugh. 1993. Pan the Tool-Maker: Investigations into the Stone Tool-Making and Tool Using Capabilities of a Bonobo (Pan Paniscus). *Journal of Archaeological Science* **20**: 81-91.

Trewavas, Anthony. 2014. *Plant Behaviour and Intelligence.* Oxford: Oxford University Press.

Trigger, Bruce G. 2003. *Understanding Early Civilizations: A Comparative Study.* Cambridge: Cambridge University Press.

Trinkaus, Erik. 1987. Bodies, Brawn, Brains and Noses: Human Ancestors and Human Predation. In *The Evolution of Human Hunting*, ed. Matthew A. Nitecki, and Doris V. Nitecki, 107-145. New York: Plenum Press.

Tucci, Serena, and Joshua M. Akey 2016. A Map of Human Wanderlust. *Nature* **538**: 179-180.

Tunnicliffe, Verena. 1991. The Biology of Hydrothermal Vents: Ecology and Evolution. *Oceanography and Marine Biology-- an Annual Review* **29**: 319-408.

Turchin, Peter. 2009. A Theory for Formation of Large Empires. *Journal of Global History* **4**: 191-217.

______ . 2011. Warfare and the Evolution of Social Complexity: A Multilevel-Selection Approach. *Structure and Dynamics* **4**(3): 185-221.

______ . 2013. The Puzzle of Ultrasociality: How Did Large-Scale, Complex Societies Evolve? In *Cultural Evolution: Society, Technology, Language, and Religion*, ed. Peter J. Richarson, and Morten H. Christiansen, 61-75. Cambridge, MA: MIT Press.

______ . 2016. *Ultrasociety: How 10,000 Years of War Made Humans the Greatest Cooperators on Earth.* Chaplin, CT: Beresta Books.

Turchin, Peter, and Sergey Gavrilets. 2009. Evolution of Complex Hierarchical Societies. *Social Evolution & History* **8**(2): 167-198.

Ungar, Peter S., and Matt Sponheimer. 2011. The Diets of Early Hominins. *Science* **334**: 190-193.

Ungar, Peter S., Frederick E. Grine, Mark F. Teaford, and Sireen El Zaatari. 2006. Dental Microwear and Diets of African Early *Homo*. *Journal of Human Evolution* **50**: 78-95.

van der Dennen, Johan M. G. 1995. *The Origin of War: The Evolution of a Male-Coalitional Reproductive Strategy*. Groningen, Netherlands: Origin Press.

______ . 1999. Human Evolution and the Origin of War: A Darwinian Heritage. In *The Darwinian Heritage and Sociobiology*, ed. Johan M. G. van der Dennen, David Smillie, and Daniel Wilson, 163-185. Westport, CT: Praeger.

Van Schaik, Carel P. 1983. Why Are Diurnal Primates Living in Groups? *Behaviour* **87**(1/2): 120-144.

Van Schaik, Carel P. and Judith Burkart. 2011. Social Learning and Evolution: The Cultural Intelligence Hypothesis. *Philosophical Transactions of the Royal Society B* **366**: 1008-1016.

Van Segbroeck, Sven, Ann Nowé, and Tom Lenaerts 2009. Stochastic Simulation of the Chemoton. *Artificial Life* **15**(2): 213-26.

Van Soest, Peter J. 1994. *Nutritional Ecology of the Ruminant* (2nd edn.). Ithaca, NY: Comstock Press.

Van Vugt, Mark. 2006. Evolutionary Origins of Leadership and Followership. *Personality and Social Psychology Review* **10**: 354-371.

Van Vugt, Mark, and Anjana Ahuja. 2011. *Naturally Selected: The Evolutionary Science of Leadership*. New York: Harper Collins.

Vasas, Vera, Chisantha Fernando, Mauro Santos, Stuart Kauffman, and Eörs Szathmáry. 2012. Evolution Before Genes. *Biology Direct* **7**: 1.

Vrba, Elizabath S., George H. Denton, Timothy C. Partridge, and Lloyd H. Burckle, eds. 1995. *Paleoclimate and Evolution, with Emphasis on Human Origins*. New Haven, CT: Yale University Press.

Vogel, Gretchen. 1998. Did the First Complex Cell Eat Hydrogen? *Science* **270**: 1633-1634.

Von Wagner, Helmuth O. 1954. Massenansammlungen von Weberknechten in Mexico. *Zeitschrift für Tierpsychologie* **11**: 349-352.

Von Uexkull, Nina, Mihai Croicu, Hanne Fjelde, and Halvard Buhaug. 2016. Civil Conflict Sensitivity to Growing-Season Drought. *Proceedings of the National Academy of Sciences* **113**: 12391-12396.

Wächtershäuser, Günter. 1988. Before Enzymes and Templates: Theory of Surface Metabolism. *Microbiology and Molecular Biology Reviews* **52**(4): 452-484.

______ . 1990. Evolution of the First Metabolic Cycles. *Proceedings of the National Academy of Sciences* **87**(1): 200-204.

______ . 1992. Groundworks for an Evolutionary Biochemistry: The Iron-sulphur World. *Progress in Biophysics and Molecular Biology* **58**(2): 85-201.

______ . 2006. From Volcanic Origins of Chemoautotrophic Life to Bacteria, Archaea and Eukarya. *Philosophical Transactions of the Royal Society B* **361**: 1787-1806.

______ . 2007. On the Chemistry and Evolution of the Pioneer Organism. *Chemistry & Biodiversity* **4**(4): 584-602.

Wächtershäuser, Günter, and Michael W. W. Adams. 1998. The Case for a Hyperthermophilic, Chemolithoautotrophic Origin of Life in an Iron-sulfur World. In *Thermophiles: The Keys to Molecular Evolution and the Origin of Life,* ed. Juergen Wiegel, 47–57. London: Taylor and Francis.

Waddington, Conrad H. 1962. *New Patterns in Genetics and Development.* New York: Columbia University Press.

Wade, Nicholas. 2006. *Before the Dawn: Recovering the Lost History of Our Ancestors.* New York: Penguin.

Wadley, Lyn, Tamaryn Hodgskiss, and Michael Grant. 2009. Implications for Complex Cognition from the Hafting of Stone Tools with Compound Adhesives in the Middle Stone Age, South Africa. *Proceedings of the National Academy of Sciences* **106**: 9590-9594.

Wallin, Ivan E. 1927. *Symbionticism and the Origin of Species.* Baltimore, MD: Williams and Wilkins.

Walsh, David M. 2015. *Organisms, Agency, and Evolution.* Cambridge, UK: Cambridge University Press.

Warren, Henry C. 2010 *Olympic: The Story Behind the Scenery.* Wickenburg, AZ: K.C. Publications, Inc.

Washburn, Sherwood L., and Chet S. Lancaster. 1968. The Evolution of Hunting. In *Man the Hunter*, ed. Richard B. Lee, and Irven DeVore, 293-303. Chicago: Aldine.

Watson, Richard A., and Eörs Szathmáry. 2016. How Can Evolution Learn? *Trends in Ecology and Evolution* **31**: 147-157.

Weber, Bruce H., and David J. Depew, eds. 2003. *Evolution and Learning: The Baldwin Effect Reconsidered.* Cambridge, MA.: MIT Press.

Weigl, Peter D., and Elinor V. Hanson. 1980. Observational Learning and the Feeding Behavior of the Red Squirrel *Tamiasciurus Hudsonicus*: The Ontogeny of Optimization. *Ecology* **61**(2): 213-218.

Weiner, Jonathan. 1994. *The Beak of the Finch.* New York: Vintage Books.

Weismann, August. 1904. *The Evolution Theory,* (trans. J. Arthur Thomson and Margaret R. Thomson). London: Edward Arnold.

Weiss, Adam. 2015. Lamarckian Illusions. *Trends in Ecology and Evolution* **30**: 566-568.

Weiss, Kenneth M., and Anne V. Buchanan. 2009. *The Mermaid's Tale: Four Billion Years of Cooperation in the Making of Living Things.* Cambridge, MA: Harvard University Press.

Weiss, Madeline C. *et al.* 2016. The Physiology and Habitat of the Last Universal Common Ancestor. *Nature Microbiology* **1**: 16116.

West, Stuart A., Ashleigh S. Griffin, Andy Gardner, and Stephen P. Diggle. 2006. Social Evolution Theory for Microorganisms. *Nature Reviews/Microbiology* **4**: 597-607.

West, Stuart A., Claire El Mouden, and Andy Gardner. 2011. Sixteen Common Misconceptions About the Evolution of Cooperation in Humans. *Evolution and Human Behavior* **32**: 231-262.

West, Stuart A., Roberta M. Fisher, Andy Gardner, and E. Toby Kiers. 2015. Major Evolutionary Transitions in Individuality. *Proceedings of the National Academy of Sciences* **112**: 10112-10119.

West-Eberhard, Mary Jane. 2003. *Developmental Plasticity and Evolution.* Oxford: Oxford University Press.

Wheeler, Peter E. 1985. The Evolution of Bipedalism and the Loss of Functional Body Hair in Hominids. *Journal of Human Evolution* **14**: 23-28.

______ . 1991. The Influence of Bipedalism on the Energy and Water Budgets of Early Hominids. *Journal of Human Evolution* **21**: 117-136.

Wheeler, William M. 1928. *The Social Insects: Their Origin and Evolution.* London: Kegan, Paul, Trench, Trubner, and Co.

White, Leslie A. 1949. *The Science of Culture: A Study of Man and Civilization.* New York: Grove Press.

______ . 1959. *The Evolution of Culture.* New York: McGraw-Hill.

White, Ronald F. 2011. Toward an Integrated Theory of Leadership. *Politics and the Life Sciences* **30**: 116-121.

White, Tim D. 1995. African Omnivores: Global Climatic Change and Plio-Pleistocene Hominids and Suids. In *Paleo Climate and Evolution: With Emphasis on Human Origins*, ed. Elizabeth S. Vrba, George Denton, Timothy Partridge, and Lloyd Burckle, 369-384. New Haven, CT: Yale University Press.

Whiten Andrew. 2011. The Scope of Culture in Chimpanzees, Humans and Ancestral Apes. *Philosophical Transactions of the Royal Society B* **366**: 997-1007.

Whiten, Andrew, and Richard W. Byrne, eds. 1997. *Machiavellian Intelligence II: Extensions and Evaluations.* Cambridge, UK: Cambridge University Press.

Whiten, Andrew, and David Erdal. 2012. The Human Socio-cognitive Niche and its Evolutionary Origins. *Philosophical Transactions of the Royal Society B* **367**: 2119-2129.

Whiten, Andrew, Robert A. Hinde, Kevin N. Laland, and Christopher B. Stringer. 2011. Introduction: Culture Evolves. *Philosophical Transactions of the Royal Society B* **366**: 938-948.

Wickramasinghe, Nalin Chandra. 1974. Formaldehyde Polymers in Interstellar Space. *Nature* **252**: 462-463.

Wilkins, Jayne, Benjamin J. Schoville, Kyle S. Brown, and Michael Chazan. 2012. Evidence for Early Hafted Hunting Technology. *Science* **338**: 942-946.

Willems, Erik P., Barbara Hellriegel, and Carel P. van Schaik. 2013. The Collective Action Problem in Primate Territory Economics. *Proceedings of the Royal Society B* **280**: 20130081.

Williams, George C. 1966. *Adaptation and Natural Selection: A Critique of Some Current Evolutionary Thought.* Princeton, NJ: Princeton University Press.

______ . 1993. Mother Nature is a Wicked Old Witch. In *Evolutionary Ethics*, ed. Matthew H. Nitecki, and Doris V. Nitecki, 217-223. Albany, NY: State University of New York Press.

Wills, Christopher. 1993. *The Runaway Brain: The Evolution of Human Uniqueness.* New York: Basic Books.

______ . 1998. *Children of Prometheus: The Accelerating Pace of Human Evolution.* Reading, MA: Perseus Books (Helix).

Wilson, David Sloan. 1997a. Introduction: Multilevel Selection Theory Comes of Age. *American Naturalist* **150**(Suppl.): S1- S4.

______. 1997b. Altruism and Organism: Disentangling the Themes of Multilevel Selection Theory. *American Naturalist* **150**(Suppl.): S122- S124.

______. 2015. *Does Altruism Exist? Culture, Genes, and the Welfare of Others.* New Haven and London: Yale University Press.

Wilson, David Sloan, and Elliott Sober. 1989. Reviving the Superorganism. *Journal of Theoretical Biology* **136**: 337-356.

Wilson, David Sloan, and Edward O. Wilson. 2007. Rethinking the Theoretical Foundation of Sociobiology. *Quarterly Review of Biology* **82**: 327-348.

Wilson, David Sloan, and Edward O. Wilson. 2008. Evolution 'For the Good of the Group.' *American Scientist* **96**: 380-389.

Wilson, Edward O. 1975. *Sociobiology: The New Synthesis.* Cambridge, MA: Harvard University Press.

Wilson, Edward O. 2013. *The Social Conquest of Earth.* New York: Liveright Publishing Corp.

Winterhalder, Bruce, and Douglas J. Kennett. 2009. Four Neglected Concepts with a Role to Play in Explaining the Origins of Agriculture. *Current Anthropology* **50**: 645-648.

Wittfogel, Karl 1957. *Oriental Despotism: A Comparative Study of Total Power.* New Haven, CT: Yale University Press.

Wolfenden, Richard, Charles A. Lewis, Jr., Yang Yuan, and Charles W. Carter, Jr. 2015. Temperature Dependence of Amino Acid Hydrophobicities. *Proceedings of the National Academy of Sciences* **112**(24): 7484-7488.

Wolpoff, Milford H. 1999a. *Paleoanthropology* (2nd edn.). New York: McGraw-Hill.

______. 1999b. The Systematics of *Homo. Science* **284**: 1774-1775.

Wolpoff, Milford H., Wu Xinzhi, and Alan G. Thorne. 1984. Modern Homo Sapiens Origins: A General Theory of Hominid Evolution Involving the Fossil Evidence from East Asia. In *The Origins of Modern Humans: A World Survey of Fossil Evidence,* eds. Fred H. Smith, and Frank Spencer, 311-327. New York: Liss.

Wolpoff, Milford H., John Hawks, and Rachel Caspari. 2000. Multiregional, Not Multiple Origins. *American Journal of Physical Anthropology* **112**: 129-136.

Wood, Bernard, and Mark Collard. 1999. The Human Genus. *Science* **284**: 65-71.

Wood, Mary Christina. 2014. *Nature's Trust: Environmental Law for a New Ecological Age.* New York: Cambridge University Press.

Woolley, Anita Williams, Christopher F. Chabris, Alex Pentland, Nada Hashmi, and Thomas W. Malone. 2010. Evidence for a Collective Intelligence Factor in the Performance of Human Groups. *Science* **330**: 686-688.

Wrangham, Richard W. 1987. The Significance of African Apes for Reconstructing Human Evolution. In *The Evolution of Human Behavior: Primate Models*, ed. Warren G. Kinzey, 51-71. Albany, NY: SUNY Press.

______. 2009. *Catching Fire: How Cooking Made Us Human.* New York: Basic Books.

Wrangham, Richard W., and Dale Peterson. 1996. *Demonic Males: Apes and the Origins of Human Violence.* Boston: Houghton Mifflin Co.

Wrangham, Richard W., and Rachel Naomi Carmody. 2010. Human Adaptation to the Control of Fire. *Evolutionary Anthropology* **19**(5): 187-199.

Wrangham, Richard W., James Holland Jones, Greg Laden, David Pilbeam, and Nancy Lou Conklin-Brittain. 1999. The Raw and the Stolen: Cooking and the Ecology of Human Origins. *Current Anthropology* **40**(5): 567- 594.

Wright, Justin P., and Clive G. Jones. 2006. The Concept of Organisms as Ecosystem Engineers Ten Years on: Progress, Limitations, and Challenges. *BioScience* **56**(3): 203-209.

Yang, Dayong, *et al.* 2013. Enhanced Transcription and Translation in Clay Hydrogel and Implications for Early Life Evolution. *Cornel Science Reports* https://cornell.app.box.com/clay

Yellen, John E., Alison S. Brooks, Els Cornelissen, Michael J. Mehlman, and Kathlyn Stewart. 1995. Middle Stone Age Worked Bone Industry from Katanda, Upper Semliki, Zaire. *Science* **268**: 553-556.

Yu, Yi *et al.* 2016. System Crash as Dynamics of Complex Networks. *Proceedings of the National Academy of Sciences* **113**: 11726-11731.

Zak, Paul J., ed. 2008. *Moral Markets: The Critical Role of Values in the Economy.* Princeton, NJ: Princeton University Press.

Zihlman, Adrienne L., and Nancy Tanner. 1978. Gathering and the Hominid Adaptation. In *Female Hierarchies*, ed. Lionel Tiger, and Heather Fowler, 163-194. Chicago: Beresford Books.

Zoetendal, Erwin G., Elaine E.Vaughan, and Willem M. de Vos. 2006. A Microbial World within Us. *Molecular and Microbiology* **59**(6): 1639-1650.

Zollikofer, Christoph P.E., *et al.* 2005. Virtual Cranial Reconstruction of *Sahelanthropus tchadensis. Nature* **434**: 755-759.

Index

Printed in the United States
By Bookmasters